生命的
所有可能：汉、英

That Benediction
is Where You Are

[印] 克里希那穆提 著
王晓霞 译

北京时代华文书局

CONTENTS

目录

PUBLISHER'S NOTE
引言

That Benediction is where you are consists of the last series of public talks that Krishnamurti gave in Bombay, in February 1985. He was to go there as usual for talks in 1986 also, but unfortunately he was unable to do so; terminal illness made him go straight from Madras to Ojai, where he died on the 17th of February.

《生命的所有可能》由克里希那穆提一九八五年二月在孟买进行的最后一系列公开讲话集结而成。他本打算一九八六年继续如常到那里进行讲话，但不幸的是他未能成行；不治之症使得他从马德拉斯直接前往欧亥，同年二月十七日，他在欧亥与世长辞。

>> Krishnamurti came to Bombay first in 1921, and gave talks between the years 1924 and 1938. After India's independence in 1947, his association with the city seems to have been almost continuous till 1985. Besides giving public talks, he held a series of discussions with small groups of friends. That was how dialogues as a form of communication started, and many of these dialogues have been compiled in books such as *Tradition and Revolution* and *Exploration into Insight*. He also addressed the staff and students of Bombay University in 1969 and the Indian Institute of Technology in 1984.

克里希那穆提一九二一年第一次来到孟买，并在一九二四到一九三八年间进行了多次讲演。一九四七年印度独立之后，他与这座城市的联系从未间断，一直持续到一九八五年。除进行公开讲演之外，他还与朋友们进行了一系列的小组讨论。这正是对话作为一种交流方式缘起的过程，其中的很多对话已经被编辑成书出版，如《传统与革命》和《探索洞见》。一九六九年和一九八四年，他也分别为孟买大学和印度理工学院的师生们进行了讲演。

Over the decades, Krishnamurti witnessed the alarming growth of Bombay from a wind-swept coastal town to an over-crowded, noisy and polluted metropolis, and he addressed these concerns in many of his talks. However, to him these social problems were but the symptoms of the deeper disorder latent in the psyche of every human being.

在过去的几十年中，孟买从一个微风吹拂的海边小镇，膨胀成为人口过度拥挤、嘈杂而又污染严重的大都市。克里希那穆提目睹了这个令人担忧的过程，他也在多次讲话中提到了这些担忧。然而在他看来，这些社会问题不过是每一个人内心深处所潜藏的混乱的外在表现。

Krishnamurti's public talks were generally held during weekends on the grounds of the J. J. School of Arts, which though located in the heart of the city, had an extensive canopy of trees. The Bombay audiences were perhaps the largest that Krishnamurti ever had anywhere in the world, especially in the 1970s and 1980s. They also represented a wide cross-section of society: scholars, intellectuals, politicians, businessmen, artists, housewives, sannyasis, students, as also Hindus, Muslims, Christians, Buddhists, Jains and Parsees.

克里希那穆提的公开讲演通常在周末位于市中心的J. J. 艺术学院的空地上举行，这里有着成片成片的树荫。孟买的听众也许是克里希那穆提在全世界拥有的最庞大的听众，特别是在二十世纪七十年代和八十年代。他们也代表着一个广泛的跨领域的社会阶层：有学者、知识分子、政客、商人、艺术家、家庭主妇、学生，也有印度教徒、穆斯林、基督教徒、佛教徒、耆那教徒和拜火教徒。

The talks in this book are remarkable for the unusual perspectives and nuances that Krishnamurti offers on the psychological issues he deals with. In the second talk, for instance, he raises various questions regarding insecurity, fragmentation, identification, and fear, but insists on the importance of merely listening to the questions and not doing anything about them. The listening, he says, is like planting a seed in the earth. "What is important is to put the question...Let the question itself answer—like the seed in the earth. Then you will see that the seed flowers and withers. Don't pull it out all the time to see if it is growing."This idea runs like a refrain throughout the talk.

本书中收录的这些讲话不同凡响，因为克里希那穆提为他所谈及的诸多心理问题，提供了非同寻常的视角和细微敏锐的洞见。例如在第二次讲话中，他就不安全、碎片化、认同和恐惧提出了各式各样的问题，但他坚持认为重要的是聆听问题，同时对它们不要做任何事情。他说，这种聆听就像是在把种子播进土地里。“重要的是提出问题……让问题自己来回答——就像播入土地的种子一样。此时你就会发现种子会成长绽放然后枯萎。不要老是把它拔出来看看它是不是在生长。”这个说法就像歌曲的副歌一样贯穿在这次讲话的始终。

There is a sense of poignancy in the substance and tone of

the last talk, where Krishnamurti urges us to realize that we are wasting our lives by not freeing ourselves from our hurts, conflicts, fears, and sorrows, and by remaining in our narrow world of specialization. This freedom, he says, is the "first step". The talk ends on a deeply religious note with his profound observation: "So, if you give your heart and mind and brain, there is something that is beyond all time. And there is the benediction of that. Not in temples, not in churches, not in mosques. All Possible Life."

最后一次讲话的内容和语气中则含有一种辛辣的意味。克里希那穆提在讲话中督促我们意识到我们在浪费自己的生命，因为我们没有把自己从伤害、冲突、恐惧和悲伤中解放出来，我们依旧停留在自己狭隘的专门化的领域之中。他说，这种解放是“第一步”。这次讲话以一句极富宗教意味的话结束，其中包含了他深邃的洞察：“所以，如果你付出你全部的身心和头脑，你就会发现有一种超越所有时间的事物，此时就会有那样一种至福。它不在庙宇中，不在教堂中，不在清真寺中，那至福就在你身边。”

Included in this book are a few excerpts from Krishnamurti’s writings which capture the beauty of Bombay’s waterfront and the atmosphere of the city, as also his sensitivity to people, the rich and the poor.

本书中还包含了来自克里希那穆提其他著作的一些节选，它们捕捉到了孟买海滨的美丽和这座城市的氛围，也体现了他对人们敏感的关怀，无论对方富贵还是贫贱。

EXCERPTS FROM WRITINGS

作品节录

The sea was very calm and there was hardly a ripple on the white sands. Around the wide bay, to the north, was the town, and to the south were palm trees, almost touching the water. Just visible beyond the bar were the first of the sharks, and beyond them the fishermen's boats, a few logs tied together with stout rope. They were making for a little village south of the palm trees. The sunset was brilliant, not where one would expect it, bit in the east; it was a countersunset, and the clouds, massive and shapely, were lit with all the colours of the spectrum. It was really quite fantastic, and almost painful to bear. The waters caught the brilliant colours and made a path of exquisite light to the horizon.

From Chapter 13 "Virtue" in *Commentaries on Living* First Series

大海非常平静，白色的沙滩上几乎没有一丝波纹。围绕着宽阔的海湾，北面是城镇，南面是棕榈树林，它们几乎触及了海面。远远的栅栏之外依稀可以看到鲨群的头领，更远处是渔夫的小船，还有几根圆木用结实的绳索绑在了一起。他们正朝棕榈树林以南的一个小村庄走去。晚霞光辉绚烂，并没有出现在人们预期的地方，而是稍稍偏向了东方，出现在相反的方向上；庞大而形状各异的云朵，被染上了七彩的光芒。这幅景象真是壮丽异常，承载起来几乎是痛苦的。海水捕捉到了绚丽的色彩，铺就了一道华美的光之路，一直延伸到海天相接之处。

摘自《生命的注释》第一卷第十三章“美德”

The sea was very calm that morning, more so than usual, for the wind from the south has ceased blowing, and before the north-easterly winds began, the sea was taking a rest. The sands were bleached by the sun and salt water, and there was a strong smell of ozone, mixed with that of seaweed. There wasn't anyone yet on the beach, and one had the sea to oneself. Large crabs, with one claw much bigger than the other, moved slowly about, watching, with the large claw waving in the air. There were also smaller crabs, the usual kind, that raced to the lapping water, or darted into round holes in the wet sand. Hundreds of sea-gulls stood about, resting and preening themselves. The rim of the sun was just coming out of the sea, and it made a golden path on the still waters. Everything seemed to be waiting for this moment—and how quickly it would pass! The sun continued to climb out of the sea, which was as quiet as a sheltered lake in some deep woods. No woods could contain these waters, they were too restless, too strong and vast; but that morning they were mild, friendly and inviting.

那天早上海面非常平静，比平时还要平静，因为来自南方的风已经停止吹送，在东北风重新吹起之前，大海得到了片刻的歇息。沙滩被烈日和咸咸的海水漂白，空气里有一股强烈的臭氧气味，与海草的味道交织在一起。海滩上还没有人，此时只有你一人与大海相对。有一些硕大的海蟹，它们的一只蟹钳远远大过另一些，它们正缓慢地四处爬动，打量着四周，大蟹钳在空中挥舞着。还有一些小螃蟹，属于常见的品种，正争先恐后地冲向拍岸的海水，或者猛冲入潮湿的沙地上那些圆圆的小孔。成百上千只海鸥站在周围，在那里休息并整理着自己的羽毛。太阳的边缘刚刚露出海面，在安静的海水上铺出一条金光大道。万物似乎都在等待这一刻——而这一刻又是那么转瞬即逝！太阳继续爬出海面，此时的大海安静得像是深林里一面得到掩蔽的湖水。没有哪片树林能够容纳这些海水，它们太过不安，太过强大和广博；然而那天早上它们却非常温和，友好而又怡人。

Under a tree above the sands and the blue water, there was going on a life independent of the crabs, the salt water and the sea-gulls. Large, black ants darted about, not making up their minds where to go. They would go up the tree, then suddenly scurry down for no apparent reason. Two or three would impatiently stop, move their heads about, and then, with a fierce burst of energy, go all over a piece of wood which they must have examined hundreds of times before; they would investigate it again with eager curiosity, and lose interest in it a second later. It was very quiet under the tree, though everything about one was very much alive. There was not a breath of air stirring among the leaves, but every leaf was abundant with the beauty and light of the morning. There was an intensity about the tree—not the terrible intensity of reaching, of succeeding, but the intensity of being complete, simple, alone and yet part of the earth. The colours of the leaves, of the few flowers, of the dark trunk, were intensified a thousandfold, and the branches seemed to sustain the heavens. It was incredibly clear, bright and alive in the shade of that single tree.

From Chapter 18 "To Change Society You Must Break Away from It" in *Commentaries on Living* Third Series

在沙滩和蔚蓝海水之上的一棵树下，独立于海蟹、盐水和海鸥之外，有一种生活在热热闹闹地进行着。大大的黑蚂蚁四处奔忙，根本无须下定决心要去往哪里。它们会爬到树上，然后突然毫无因由地匆忙赶下来。有两三只会不耐烦地停下来，脑袋四处张望着，然后带着一股喷涌而出的劲头，绕着一小块木片团团转起来。它们以前必定已经检查过那块木片成百上千次了；它们会带着迫切的好奇心再把它审视一遍，然后在一秒钟之后就对它失去了兴趣。树下非常安静，尽管你周围的一切都异常活跃。树叶间没有一丝微风扰动，但每片树叶都充盈着美和晨光。那棵树身上有一股热切——不是求取、成功那类糟糕的热切，而是完满、简单、孑然独立而又作为大地一部分的那种热切。浓密的树叶、稀疏的花朵和黝黯树干的色彩，成千倍地浓郁起来，茂盛的树枝像是在支撑着天空。在这一棵树的树荫之下，天空令人难以置信地清澈、明亮和生机勃勃。

016

摘自《生命的注释》第三卷第十八章“若要改变社会，你就必须从中脱离出来”

It was hot and humid and the noise of the very large town filled the air. The breeze from the sea was warm, and there was the smell of tar and petrol. With the setting of the sun, red in the distant waters, it was still unyieldingly hot. The large group that filled the room presently left, and we went out into the street.

天气炎热潮湿，大城市的噪音充斥在空气之中。来自大海的微风也是温热的，带着一股焦油和汽油的味道。炽热的红日落入了遥远的海面，但天气依然不屈不挠地炎热着。之前满满一屋子的人一下子都离开了，我们也走出房间来到了街道上。

The parrots, like bright green flashes of light, were coming home to roost. Early in the morning they flew to the north, where there were orchards, green fields and open country, and in the evening they came back to pass the night in the trees of the city. Their flight was never smooth but always reckless, noisy and brilliant. They never flew straight like other birds, but were forever veering off to the left or the right, or suddenly dropping into a tree. They were the most restless birds in flight, but how beautiful they were with their red beaks and a golden green that was the very glory of light. The vultures, heavy and ugly, circled and settled down for the night on the palm trees.

鹦鹉们像是明亮的绿色闪电，正在回巢栖息。清晨它们早早地飞向北方，那里有果园、绿色的田野和开阔的乡间，到了傍晚它们又飞回来，在城里的树上过夜。它们的飞翔从来都不平缓，而是始终都粗心、嘈杂而又耀眼。它们从来不像其他鸟儿那样直直地飞行，而是永远都左突右闪地改变方向，或者突然落在一棵树上。它们是在飞行中最不安分的鸟儿，然而它们又是那么美丽，有着红色的尖喙和金光闪闪的一袭绿衣，那色彩正是光的荣耀。而那些秃鹫笨重又丑陋，盘旋着落下，然后夜晚就栖息在棕榈树丛上。

A man came along playing the flute; he was a servant of some kind. He walked up the hill, still playing, and we followed him; he turned into one of the side-streets, never ceasing to play. It was strange to hear the song of the flute in a noisy city, and its sound penetrated deep into the heart. It was very beautiful, and we followed the flute player for some distance. We crossed several streets and came to a wider one, better lighted. Farther on, a group of people were sitting cross-legged at the side of the road, and the flute player joined them. So did we; and we all sat around while he played. They were mostly chauffeurs, servants, night watchmen, with several children and a dog or two. Cars passed by, one driven by a chauffeur; a lady was inside, beautifully dressed and alone, with the inside light on. Another car drew up; the chauffeur got out and sat down with us. They were all talking and enjoying themselves, laughing and gesticulating, but the song of the flute never wavered, and there was delight.

有个人吹着笛子一路走来，他有着某种仆从的身份。他爬上山去，依然吹奏着，我们则跟在他后面；然后他转入了一条小巷，但从来没有停止吹奏。在一座嘈杂的城市里听到悠扬的笛声，真是一件奇妙的事情，它的声音摄人心魄。那笛声非常优美，我们跟随着吹笛人走了一段路程，然后穿过几条街来到了一条宽阔、灯光也明亮一些的街道上。更远处有一群人正盘腿坐在路边，那个吹笛人加入了他们，我们也加入了进去。他吹笛子的时候我们都坐在他的周围。这里的人多数是司机、仆人和守夜者，有着几个孩子和一两条狗。有汽车从旁边驶过，其中一辆由司机驾驶，车里的灯亮着，一位女士独自坐在车上，衣着靓丽。另一辆车开过来，司机下了车，和我们坐在一起。他们都在聊着天，大笑着，用手比画着，自得其乐，而那笛声从未停息过，有一种欢欣弥漫在四周。

Presently we left and took a road that led to the sea past the well-lit houses of the rich. The rich have a peculiar atmosphere of their own. However cultured, unobtrusive, ancient and polished, the rich have an impenetrable and assured aloofness, that inviolable certainty and hardness that is difficult to break down. They are not the possessors of wealth, but are possessed by wealth, which is worse than death. Their conceit is philanthropy; they think they are trustees of their wealth; they have charities, create endowments; they are the makers, the builders, the givers. They build churches, temples, but their god is the god of their gold. With so much poverty and degradation, one must have a very thick skin to be rich. Some of them come to question, to argue, to find reality. For the rich as for the poor, it is extremely difficult to find reality. The poor crave to be rich and powerful, and the rich are already caught in the net of their own action; and yet they believe and venture near. They speculate, not only upon the market, but upon the ultimate. They play with both, but are successful only with what is in their hearts. Their beliefs and ceremonies, their hopes and fears have nothing to do with reality, for their hearts are empty. The greater the outward show the greater the inward poverty.

From Chapter 7 " The Rich and the Poor " in *Commentaries on Living* First Series

不久我们就离开了人群，走上了一条通往海边的路，路旁经过的是富人们灯火通明的宅邸。富人们身上有着一股他们特有的气息。无论多么有文化、多么谦逊、多么有古风、多么优雅，他们总是带着一股顽固的、骄傲的冷漠态度，那是一种很难打破的不可侵犯的确定和坚硬。他们并非财富的占有者，而是被财富所占有，可这比死亡还要糟糕。他们的自负来自他们的慈善活动，他们自以为是自己财富的掌管者；他们拥有慈善事业，四处捐赠；他们是缔造者、修建者、施予者。他们建造教堂、庙宇，而他们的神明是他们的财富之神。世上有着如此严重的贫穷和腐化，一个人必须厚颜无耻才能致富。他们之中有一些人前来质疑、争辩和寻找真相。对于富人来说，发现真相是极其困难的事情，就和穷人一样。穷人渴望得到财富和权势，而富人已经困在了自身行为的罗网中，但他们却依然抱持着信仰，依然趋近涉险。他们不仅在市场上投机，而且投机于那个终极之物。他们对两者都加以玩弄，但成功收获的只是他们内心的现状。他们的信仰和仪式，他们的希望和恐惧，都与真相无关，因为他们内心是空洞的。外在越是强大，说明内心就越贫乏。

摘自《生命的注释》第一卷第七章“富人与穷人”

TALK Ⅰ 讲话一

思考是记忆的运动

THIS is a dialogue between us, a conversation between two friends. So this is not a lecture to instruct, inform or guide you. We are going to talk over together many things, certainly not to convince you of anything, or to inform you of new ideas, new concepts, conclusions or ideals. We are going to look together at the whole world as it is, at what is happening not only in this part of the world but also in the rest of the world. Together. And the speaker means together. You and he are going to observe, without any bias, without any prejudice, what is happening globally.

这是我们之间的一场对话，是两个朋友间的一场交谈，所以这不是一场用来指导你或者向你灌输的讲座。我们要一起来探讨某些事情，而绝不是要说服你相信什么，也不是向你灌输新想法、新观念、新结论或新理想。我们要一起来看看整个世界的现状，不仅看看世界的这个部分，而且也看看世界的其他地方发生着什么。一起来看看。讲话者说的是“一起”。你和他一起来观察全球正在发生的事情，不带有任何先入之见，也不带有任何偏见。

So this is a serious talk, not something intellectual, emotional or devotional. So we must exercise our brains. We must have scepticism, doubt; we must question and not accept anything that anybody says—including all your gurus and sacred books. We have come to a crisis in the world. The crisis is not merely economic; rather it is psychological. We have lived on this earth for over millions of years and, during that long period of time, we have passed through every kind of catastrophe, every kind of war. Civilizations have disappeared; so have cultures that shaped the behaviour of human beings. We have had a great many leaders, political and religious, with all the tricks that they have played on human beings. And after this enormous evolution of the human brain, we are what we have been—rather primitive, barbarous, cruel, and always preparing for war. Every nation now is storing up armaments. And we human beings are caught in this wheel of time. We have not changed very much; we are still barbarians, with all kinds of superstitions and beliefs. At the end of it all, where are we?

所以说这是一场严肃的讲话，而不是一件智力上、情感上或者信仰上的事情，因此我们必须运用我们的大脑。我们必须抱有怀疑和质疑精神，必须质疑而不是接受任何人说的任何事情——包括你所有的古鲁和圣典。一场世界性的危机已经来到我们面前，这场危机不仅仅是经济上的，而且是心理上的。我们已经在这个地球上生活了数百万年，在时间的长河中，我们历经了各种各样的灾难，各种各样的战争。无数文明曾经消失过，塑造了人类行为的无数文化也遭受过同样的命运。我们有过无数的政治或宗教领袖，他们都在人类身上玩过不计其数的把戏。而在人类的大脑经过了这番非同寻常的进化之后，我们还是过去的那副样子——原始、野蛮、残忍，总是在为战争做准备。每个国家都在囤积军备。而我们人类就被困在了时间的车轮之中。我们从来都没有改变多少，我们依然是野蛮人，有着各式各样的迷信和信仰。在这一切的最后，我们又到了哪里呢？

Please, we are talking over things together. It is not that the speaker is explaining all this; it is so obvious. You and the speaker are together examining very carefully and diligently, what we have become and what we are. And we ask: Will time change us? Will time, that is, another fifty thousand or a million years change the human mind, the human brain? Or is time not important at all? We are going to talk about all these things.

请注意，我们是在一起探讨问题，并不是讲话者在解释一切，这一点再明显不过了。你和讲话者是在一起非常仔细、孜孜不倦地审视我们变成了什么样子，我们现在究竟如何。而我们问：时间会改变我们吗？时间，也就是再花上五万年或者一百万年，会改变人类的心智、人类的大脑吗？我们要来一起探讨所有这些事情。

Human beings are wounded psychologically. Human beings throughout the world are caught in great sorrow, pain, suffering, loneliness and despair. And the brain has created the most extraordinary things, ideologically, technologically, religiously. The brain is extraordinarily capable. But that capacity is very limited. Technologically we are advancing at an extraordinary speed. But psychologically, inwardly, we are very primitive, barbarous, cruel, thoughtless, careless, and indifferent to what is happening. We are indifferent not only to the corruption that goes on environmentally but also to the corruption that goes on in the name of religion, in the name of politics, business, and so on. Corruption is not just passing money under the table or smuggling goods into the country. Corruption begins where there is self-interest. Where there is self-interest, that is the origin of corruption.

人类在心理上备受伤害，全世界的人类都困在了巨大的悲伤、痛苦、不幸、孤独和绝望之中。大脑在意识形态、科技和宗教方面都成就了最为非凡的事物。大脑具有非同寻常的能力，但这种能力是非常局限的。我们在技术上以飞快的速度取得进步，但是在心理上、在内心，我们依然非常原始、野蛮、残忍、自私、冷漠，对世上发生的事情无动于衷。我们不仅仅对于环境中发生的腐败无动于衷，而且对以宗教之名、以政治和商业之名进行的腐败也无动于衷。腐败不仅仅是行贿受贿或者往国内走私物品，而是只要存在自私自利，腐败就会产生。存在自私自利的地方，那就是腐败的发源地。

Are we thinking together, or are you merely listening to the speaker? Are we going together as two friends, taking a long journey—a journey into the global world, a journey into ourselves: into what we are, what we have become, and why we have become what we are. And we need to take this journey together. It is not that the speaker takes the journey and points out to you the map, the road, and the way. But, rather, we are together, and the speaker means together. For he is not a guru. One should not follow anybody in the world of thought, in the world of the psyche. We have depended so much on others to help us. And we are not helping you. Let us be very clear on that point: the speaker is not helping you because you have had helpers galore. And we have not been able to stand alone, think out things for ourselves; we have not been able to look at the world and our relationship to the world, and see whether we are individuals at all or part of humanity. We have not exercised our brains, which are so extraordinarily capable. We have expended our energy, our capacity, our intellectual understanding in one direction only—the technological. But we have never understood human behaviour and why we are as we are after this long period of evolution.

我们是在一起思考，还是说你只不过是在听讲话者说话？我们是在像两个朋友那样一起踏上一段长长的旅程吗？——一段进入整个世界的旅程，一段深入我们内心的旅程：深入探究我们实际的样子、我们已然如何，而且为什么我们会成为现在的样子。我们需要一起踏上这段旅程。并不是讲话者踏上旅程，然后把地图、路径和方法指给你看，而是我们要并肩而行，并且讲话者说的就是“一起”，因为他不是一个古鲁。在思想世界中，在精神世界中，你不应该追随任何人。我们已经太依赖别人来帮助我们了，而现在我们并不是在帮助你。让我们把这一点明确：讲话者并不是在帮助你，因为你已经有太多的帮助者了。我们一直未能独立，自己把事情想清楚；我们一直没有能力去看这个世界以及我们与世界的关系，一直未能看清我们究竟是个体还是只是人类的一部分。我们没有运用我们的大脑，这能力非凡的大脑。我们只把自己的能量、能力和智力上的理解力运用在某一个方向——技术层面。但我们从来未曾了解人类的行为，以及为什么在漫长的进化之后我们变成了现在这个样子。

And as the speaker said just now, he is not helping you; we are together looking, understanding. Of course, we need the help of a physician or a surgeon. We depend on governments, however rotten they are. We have to depend on the postman, and the milkman, and so on. But to ask for help through prayer, through meditation seems so utterly futile. We have had such help; we have had thousands of gurus and thousands of books—so-called religious and non-religious. And in spite of them all we are helpless. We may earn a lot of money, have big houses, cars, and so on, but psychologically, inwardly, subjectively, we are almost helpless because we have depended on other people to tell us what to do, what to think. So, please, the speaker is saying most respectfully, seriously, and earnestly that he is not trying to help you. On the contrary, we are together.

而正如讲话者刚刚所说的那样，他并不是在帮助你；我们是在一起去看，去了解。当然，我们需要内科医生或者外科医生的帮助。我们依赖于政府，无论它们有多么腐败。我们也得依靠邮差和送奶工等。但是通过祈祷、通过冥想来寻求帮助，看起来是那么徒劳无益。我们得到过这样的帮助，我们已经有了数千个古鲁和数千本书——既包括所谓宗教的，也包括非宗教的。尽管它们都存在，但我们依旧茫然无助。我们也许赚了很多钱，有了大房子、车子等，但从心理上、从内在、从主观上我们几乎是孤立无援的，因为我们依赖别人来告诉我们该做什么、想什么。所以请注意，讲话者是抱着极大的敬意，用极其认真、极其恳切的态度表明：他并不想帮助你。相反，我们是在一起的。

So you and the speaker have to investigate all this: our relationship to the world, which is becoming more and more complex, our relationship to each other however intimate it might be, our relationship to an ideal, our relationship to our gurus, and to so-called God. We have to inquire seriously, deeply, into the quality of a brain that comprehends, or has an insight into the whole outer as well as the psychological world in which we live. It must be clear that we are not trying to point out a way, a method, a system, or in any way trying to help you. On the contrary, we are independent human beings. This is not a cruel or indifferent statement. We are like two friends talking over together, trying to understand the world: the environment, all the complications of the economic world, the separate religions, and separate nations. Friendship means that we are not trying to persuade, coerce or impress each other. We are friends and, therefore, there is a certain quality of affection, understanding, exchange. We are in that position.

所以你和讲话者必须一起来探究这一切：我们与世界的关系，这关系正变得越来越复杂，我们与彼此的关系，无论它有多么亲密，还有我们与理想的关系，我们与我们的古鲁以及所谓“神”的关系。我们必须认真地、深入地探询能够领悟或者洞察我们所处的整个外部及内心世界的头脑所具有的品质。有一点必须清楚，那就是：我们不想指出任何一种道路、方法或者体系，也绝不试图帮助你。恰恰相反，我们是互相独立的人。这并不是一个残忍或者冷漠的说法。我们就像两个朋友一样在一起探讨，试着理解这个世界：环境，经济世界的所有复杂之处，分裂的各种宗教，还有分裂的各个国家。朋友就意味着我们不试图说服、强迫或者影响对方。我们是朋友，所以我们之间有一种友爱、理解和交流的品质。我们就处在这样的位置上。

So we first begin with what our brain is. The speaker is not a brain specialist, but he has talked with brain specialists. The brain, which is inside the skull, is a most extraordinary instrument. It has acquired tremendous knowledge about almost everything. It has invented the most incredible things like the computer, the means of quick communication, and instruments of war. And here it is entirely free to investigate, invent, research. It starts with knowledge, and accumulates more and more knowledge. If a certain theory does not work, it is dropped. But the brain is not equally free to inquire into the self. It is conditioned, shaped, programmed—to be a Hindu, a Muslim, a Christian, a Buddhist, and so on. Like a computer, the human brain is programmed—that you must have war, that you belong to a certain group, that your roots are in this part of the world, and so on. This is correct; this is not an exaggeration. All of us are programmed by tradition, by the constant repetition in newspapers and magazines, by thousands of years of pressure. The brain is free in one direction: in the world of technology. But that very brain, which is so extraordinarily capable, is limited by its own self-interest. Our brain, which is amazingly free in one direction, is psychologically a cripple.

那么我们首先从探讨我们的大脑是什么开始。讲话者不是一个大脑专家，但他跟一些大脑专家交谈过。头颅内的大脑，是一个极其非凡的工具，它获得了几乎关于一切的海量知识。它发明了最不可思议的东西，比如计算机、快捷的通信手段及各种战争武器。在这些方面它可以完全自由地探索、发明、研究。它从知识出发，然后累积了越来越多的知识。如果某种理论行不通，它就会将其放弃。但大脑在探究自我方面就没有那么自由了。它被制约、塑造和程式化了——成了一名印度教徒、穆斯林、基督教徒、佛教徒等。就像计算机一样，人类的大脑被程式化了——你必须发动战争，你要属于某个团体，你的根就扎在世界的这个地方，等等。这是事实，并非夸张。我们所有人都被传统所程式化，被报纸杂志不停地重复宣传、被几千年来的压力所程式化。大脑在某个方向上——在科技世界中是自由的。但正是这个大脑，这个极其能干的大脑，被它自身的自私所局限了。我们的大脑，在某个方向上出奇地自由，在心理上却严重残疾。

Is it possible for the human brain to be entirely free so that there is tremendous energy? Not to do more mischief, not to have more money, or power—though you must have money—but to inquire, to find out a way of life in which there is no fear, no loneliness, and no sorrow, and to inquire into the nature of death, meditation and truth. Is it possible for the human brain, which has been conditioned for thousands of years, to be entirely free? Or must human beings everlastingly be slaves, never knowing what freedom is — not freedom in the abstract but freedom from conflict, because we live in conflict. One fact common to all human beings—from childhood till they die—is this constant struggle, seeking security and therefore never finding it, or being insecure, wanting security. So is it possible for human beings in the modern world with all its complexities to live without a shadow of conflict? Because conflict distorts the brain, lessens its capacity, its energy, and the brain soon wears itself out. You can observe in yourself as you grow older this perpetual conflict.

人类的大脑有可能完全自由，因而拥有惊人的能量吗？——不是为了制造更多的伤害，不是为了获取更多的金钱或权力——尽管你必须有钱——而是为了去探究，为了找到一条没有恐惧、没有孤独、没有悲伤的生活之道，为了探究死亡、冥想和真理的本质。人类的大脑已经被制约了数千年，它有可能彻底自由吗？还是说，人类永远都是奴隶，永远都不会知道自由是什么——不是抽象的自由，而是摆脱了冲突的自由，因为我们就生活在冲突之中。对所有人类来说有一个共同的事实，那就是这种不断的冲突从孩童时代一直持续到他们死去的那一刻——寻求保障却因此从来都找不到，或者感觉不安全于是追求安全感。那么，当今这个极其复杂的世界中的人类，有可能活得没有一丝冲突的阴影吗？因为冲突会扭曲大脑，削弱它的能力和能量，于是大脑很快就会消耗殆尽。随着你年纪渐长，你从自己身上就可以观察到这种永无止境的冲突。

What is conflict? Please do not wait for me to answer it; that is no fun at all. Ask yourself that question, and give your mind to find out what is the nature of conflict. Conflict exists, surely, when there is duality:“me”and“you”, my wife separate from me, the division between the meditator and meditation. So, as long as there is division between nationalities, between religions, between people, between the ideal and the fact, between “what is”and“what should be”, there must be conflict. That is a law. Wherever there is separation, the sense of division as the Arab and the Jew, the Hindu and the Muslim, the son and the father, and so on, there must be conflict. That is a fact. That division is the“more”: “I do not know, but give me a few years and I will know.”I hope you understand all this.

冲突是什么？请不要等我来回答这个问题，那样一点都不好玩儿。问问你自己这个问题，用心去弄清楚冲突的本质是什么。当二元性存在时：“我”和“你”，我妻子和我是分开的，冥想者和冥想之间是分离的，那么冲突就必然会存在。所以，只要存在各个国家、各派宗教、各种人群之间的划分，只要存在理想与现实、“现在如何”与“应当如何”之间的划分，就必然会存在冲突。这是一条铁律。哪里有分离，哪里有阿拉伯人与犹太人、印度教徒与穆斯林、父与子等之间的划分，哪里就必然会有冲突。这是一个事实。这种划分就是想要“更多”：“我不知道，但是给我几年时间我会知道的。”我希望这些你都明白了。

Who has created this division between“what is”and“what should be”? There is the division between so-called God, if there is such an entity, and yourself, and the division between wanting peace and being in conflict. This is the actual reality of our daily life. And the speaker is asking, as you must be asking too: who has created this division, not only externally but also inwardly? Please ask yourselves this question. Who is responsible for all this mess, this endless struggle, endless pain, loneliness, despair, and a sense of sorrow from which man seems to have never escaped? Who is responsible for all this? Who is responsible for the society in which we live? There is immense poverty in this country. Do you understand all this, or you have never thought about it at all? Or are you so occupied with your own meditations, with your own gods, with your own problems, that you have never looked at all this, never asked?

那么，是谁制造了“现在如何”与“应当如何”之间的这种划分？所谓的“神”——如果有这种存在体的话——与你自己之间存在着划分，渴望和平与身处冲突之间也存在着分别。这就是我们日常生活中真切的现实。而讲话者在问，你必定也在问：是谁从外在和内在制造了这种划分？请问问你们自己这个问题。谁对这所有的混乱，这无尽的挣扎、无尽的痛苦、孤独、绝望，以及人类似乎从未逃脱的悲伤感负责？谁对这一切负责？谁对我们所生活的这个社会负责？这个国家有着无尽的贫穷。你明白这些吗？还是说你根本就没有想过这个问题？或者你满脑子都被自己的冥想、自己的神明、自己的问题所占据，以至于从来都没有看过这些事实，从来都没有问过这些问题？

There are several things involved in all this. Those who are fairly intelligent, fairly aware, and sensitive have always sought an egalitarian society. They have asked: Can there be equal opportunity, no class difference, so that there is no division between the worker and the manager, the carpenter and the politician? So we ask: is there justice in the world? There have been revolutions like French revolution—which has tried to establish a society where there is equality, justice, and goodness. But they have not succeeded at all. On the contrary, they have gone back to the old pattern in a different setting. So you have to inquire not only into why human beings live in perpetual conflict and sorrow, and why they search for security, but also into the nature of justice. Is there any justice at all in the world? Is there? You are clever, another is not. You have got all the privileges, and another has none whatsoever. You live in a palatial house, and another lives in a hut, having hardly one meal a day. So is there justice at all? Is it not important to find out for oneself and, therefore, help humanity.(I am sorry, I do not mean“help”; I withdraw that word.)To understand the nature of justice and find out if there is any justice at all, you must inquire very, very deeply into the nature of sorrow, and whether there can be no self-interest at all. And also we should inquire into what is freedom and what is goodness.

其中涉及几件事情。那些相当明智、相当觉知和敏感的人，总是在寻求一个人人平等的社会。他们曾经问过：能不能有平等的机会，没有阶级差异，因而也没有工人和经理、木匠和政客之间的分别？所以我们问：世界上存在公平吗？无数次革命都已经发生过了——比如法国大革命曾试图建立一个公平、公正和善良的社会，但它根本没有成功。恰恰相反，它又回到了旧有的模式中，只是换了一个环境背景而已。所以你不仅需要探询为什么人类生活在无休止的冲突和悲伤中，为什么他们要追求安全，而且还要探究公平的本质。世界上究竟有任何公平存在吗？有吗？你很聪明，而另一个人不聪明。你拥有所有的特权，而另一个人一无所有。你住在宫殿般的房子里，而另一人住在草房里，一天难得有一餐饭。所以说公平到底存在吗？你亲自弄清楚这一点，进而帮助全人类，这难道不重要吗？（抱歉，我的意思不是“帮助”，我收回这个词。）若要了解公平的本质，弄清楚公平究竟是否存在，你就必须非常非常深入地探究悲伤的本质，以及有没有可能根本没有自私自利。也许我们还应该探询自由是什么、善是什么。

The society in which we live is created by every human being through his greed, envy, aggression, and the search for security. We have created the society in which we live, and then we become slaves to that society. Do you understand all this? We human beings out of fear, out of loneliness, and in our search for security—never understanding what insecurity is but always wanting security—have created our culture, our society, our religions, our gods. To come back: who has created this division? Because where there is division, there is conflict. That is an absolute certainty. Think it out, sir. Is it not thought that has divided the world as the Christians, the Buddhists, the Hindus and the Muslims? Is it not thought?

我们所生活的这个社会，由每一个人用他自己的贪婪、嫉妒、攻击性和对安全的追求所筑造。我们建造了我们所处的这个社会，然后成了它的奴隶。这些你都明白吗？我们人类出于恐惧、出于孤独，通过我们对安全的追求——从来都不了解不安全是什么，就只会不停地追求安全——造就了我们的文化、我们的社会、我们的宗教、我们的神明。我们回过头来问：是谁造成了这种划分？因为哪里有划分，哪里就会有冲突。这是绝对确定无疑的事情，先生，请好好想一想，难道不是思想把世界划分成了基督教徒、佛教徒、印度教徒和穆斯林吗？不就是思想吗？

Then we ask: what is thought? Thought is the action by which we live. Thought is our central factor of action. Right? Thought—by which we make money; thought which separates me and you, the husband and the wife, the ideal and"what is". Then, what is thought? What is thinking? Is not thinking the activity of memory? Please, sirs, do not accept a thing that the speaker is saying. You must have this quality of doubt; doubt your own experiences, your own ideas. The speaker is saying as a friend—to whom you can listen or not listen as you please—that thought has created this division. Thought has been responsible for wars, has been responsible for all the gods that man has invented. Thought has been responsible for putting man on the moon, for creating the computer and all the extraordinary things of the technological world. And thought is also responsible for the division and conflict between"what is"and"what should be". The"what should be"is the ideal; it is something to be achieved, something to be gained, away from"what is". For example, human beings are violent. That is an obvious fact. Even after a long period of time, man is not free of violence. But he has invented non-violence. He has invented it and is pursuing it. He would acknowledge that he is violent—if he is at all honest. But in the pursuit of the ideal called non-violence, he is sowing the seeds of violence all the time.

Naturally. This is a fact.

接下来我们问：思想是什么？思想是我们赖以生存的行动，思想是我们行动的核心因素，对吗？思想——我们通过它来赚钱，思想分离了我和你、丈夫和妻子、理想和“现状”。那么，思想是什么？思考是什么？思考不就是记忆的活动吗？先生们，请不要接受讲话者说的任何一件事。你必须具有这种质疑的品质，质疑你自己的经验、你自己的想法。讲话者是作为一位朋友在告诉你——想不想听都随你——是思想造成了这种划分。思想对战争负有责任，对人类发明的所有神明负有责任。思想对人类登上月球负责，对制造出计算机和科技界所有了不起的东西负责。而思想也对“现在如何”与“应当如何”之间的分裂和冲突负责。“应当如何”是理想，是某种要去实现的东西，某种要去获取的东西，它远离了“现在如何”。比如说，人类是暴力的，这是一个显而易见的事实。即使经过了漫长的时间，人类还是没有摆脱暴力。但是他发明了非暴力，把它虚构出来然后再去追求它。他会承认自己是暴力的——如果他还算诚实的话。但是，在追求所谓的“非暴力”这个理想的过程中，他却一直在播撒暴力的种子。

这自然是一个事实。

This country has talked a great deal about nonviolence. This is rather shameful because we are all violent people. Violence is not merely physical; it is also imitation, conformity, moving away from“what is”. So violence can end completely in the human mind, the human heart, only when there is no opposite. The opposite is the non-violence which is not real; it is another escape from violence. If you do not escape, then there is only violence. But you have not been able to face that fact. You are always running away from the fact, finding excuses, finding economic reasons, finding innumerable methods to overcome it, but there is still violence. The very overcoming is a part of violence.

关于非暴力，这个国家已经说得太多了。这真让人觉得羞愧，因为我们都是暴力的人。暴力不仅仅是身体上的，它也是模仿、遵从以及逃离“现在如何”。所以，只有当对立面不存在时，人类头脑和内心之中的暴力才能彻底终结。对立面就是那个不真实的非暴力，它是对暴力的另一种逃避。如果你不逃避，那么存在的就只有暴力。然而你却没能面对这个事实，你总是从事实那里逃走，寻找借口，寻找经济上的原因，寻找无数的方法来克服它，但暴力依然存在。这种克服本身就是暴力的一部分。

So, to face violence, you must give attention to it, and not run away from it. You must see what it is, see the violence between man and woman, sexually and in other ways. Is there not violence when you are seeking more and more, “becoming”more and more? So to look at violence and remain with it; do not run away from it, or try to suppress it, or transcend it—all that implies conflict. To live with it, look at it, in fact, treasure it, and not translate it according to your want, likes and dislikes. Just to look and observe with great attention. When you give attention to something completely, it is like turning on a bright light, and then you see all the qualities, the subtleties, the implications—the whole world—of violence. When you see something very clearly, it is gone. But you refuse to see things clearly.

所以，若要面对暴力，你就必须关注它，而不是逃离它。你必须看清它是什么，看到男人与女人之间在性及其他方面的暴力。当你追求更多，想要“变得”更加如何时，难道没有暴力吗？所以要看着暴力并与它共处，不要逃离它，也不要试图压抑它或者超越它——那些都意味着冲突。与它共存，实实在在地看着它，珍视它，而不是根据你的愿望、你的好恶来诠释它。只是带着巨大的关注看着它、观察它。当你对某件事情付出全部的注意力，那就像是打开了一盏明亮的灯，于是你就可以看到暴力所有的品质、所有的微妙之处、所有的含义——暴力的全部。当你非常清楚地看到了某件事情，它就消失了。但是你拒绝清晰地看到事物。

So we are asking: who has created this conflict of human beings with each other, with the environment, with the gods, with everything? Have you ever considered why you think you are an individual? Are you an individual? Or have you been programmed to think you are an individual? Your consciousness is like every other human being's consciousness. You suffer, you are lonely, you are afraid, and you seek pleasure and avoid pain. It is so with every human being on this earth. This is a fact, a psychological fact. You may be tall, you may be dark, you may be fair, but those are all external frills, of climate, food, and so on. And culture too is external. But psychologically and subjectively, our consciousness is common, one with all other human beings. You may not like it, but that is a fact. So psychologically you are not separate from the rest of humanity. You are humanity. Do not say"Yes"; it has no meaning merely accepting it as an idea. It is a tremendous fact that you are the rest of mankind, and not somebody separate. You may have a better brain, more wealth, more cunning, better looks. But put aside all that for they are all surface things, they are frills. Inwardly, every human being on this earth is one with you in sorrow. Do you realize what that means? It implies that when you say you are the rest of humanity, you have tremendous responsibility. It implies that you have great affection, love, compassion, and not some silly idea that"we are all one".

所以我们在问：是谁制造了人类彼此之间的这种冲突，以及与环境、与神明、与一切的冲突？你可曾思考过你为什么认为自己是一个个体？你是一个个体吗？还是你被程式化了于是认为自己是一个个体？你的意识就跟其他任何一个人的意识一样，你受苦，你孤独，你恐惧，你寻求快乐、避开痛苦。这个地球上的每个人都是如此。这是一个事实，一个心理事实。你也许身材高大，你也许是黑皮肤，你也许是白皮肤，但这些都是气候、饮食等外在的虚饰。而文化也是外在的。但是从心理上、从主观上，我们的意识都是相同的，与其他所有人类都共有同一个意识。你也许不喜欢这一点，但这是事实。所以你和其他人类在心理上并不是分开的，你就是人类。不要说“是的”；单纯把它当作一个理念来接受，没有任何意义。你就是其他人类，而不是一个分离出来的人，这是一个无比非凡的事实。你也许有颗更好用的头脑，你也许更富有、更精明、更漂亮。但是这一切都要抛在一边，因为它们都是表面上的事情，它们都是无用的虚饰。从内在来讲，这个地球上的每一个人都和你是一体的，都身处悲伤之中。你明白这意味着什么吗？这意味着，当你说你就是其他人类的，你就有了惊人的责任感。这意味着你有了巨大的关怀、爱和慈悲，而不是一个愚蠢的想法——“我们都是一体的”。

So we must inquire into what is thinking and why thinking has become so extraordinarily important. Thinking cannot exist without memory. If there was no memory, there would be no thought. Our brains—which are one with all the rest of humanity, and not separate little brains—are conditioned by knowledge, by memory. And knowledge, memory, is based on experience, both in the scientific world and the subjective world. Our experiences, however subtle, however spiritual, and however personal, are always limited. So our knowledge, which is the outcome of experience, is also limited. And we are adding more and more; and where there is addition, that which is being added to is limited. So we are saying that experience being limited, knowledge is always limited, either now, in the past, or in the future.

所以我们必须探究思想是什么，为什么思想变得非同寻常地重要。没有记忆，思想就无法存在。如果没有记忆，就不会有思想。我们的大脑——我们与其他所有人类拥有的是同一个大脑，而不是一个个分离开来的小大脑——被知识和记忆所制约。而知识和记忆以经验为基础，既包括科学世界的经验，也包括主观世界的经验。我们的经验，无论多么微妙、多么崇高、多么个人化，都始终是局限的。所以我们的知识，也就是经验的产物，也是局限的。并且我们一直在添加越来越多的知识，而只要存在添加，那些被添加的东西就是局限的。所以我们说经验是局限的，因此知识也始终是局限的，无论现在、过去还是未来。

And knowledge means memory—the memory held either in the computer or in the brain. So the brain is memory. And that memory directs thought. This is a fact. So thought is always limited. Right? This is logical, rational, not something invented; this is so. Experience is limited. Therefore knowledge is limited.

知识就意味着记忆——这记忆要么储存在电脑里，要么储存在大脑里。所以大脑就是记忆，而这记忆会支配思想，这是一个事实。所以思想始终是局限的，对吗？这是符合逻辑的、合理的，不是捏造出来的东西，事实就是如此。经验是局限的，因而知识也是局限的。

Then we ask: Is there any other activity which is not divisive, which is not fragmentary, which does not break up? Is there a holistic activity that can never break up—as me and you? It is division which creates conflict. Now, how are you going to find this out for yourselves, seeing that thought is divisive, that thought creates conflict, that thought has created the society and then set you apart from the society which you have created? Thought is the only instrument we have had so far. You may say that there is another instrument which is intuition. But that can be irrational; you can invent anything and live in an illusion.

然后我们问：有没有另外一种行动，它不是分裂的，不是支离破碎的，也不会分崩离析？有没有一种完整的行动，它绝不会分裂——分裂成我和你？正是分裂造成了冲突。那么，看到了思想具有分裂性，思想会制造冲突，思想建立了这个社会，然后又把你从自己建立的社会中分离出来，那么接下来你要如何亲自找到这种行动呢？思想是目前我们拥有的唯一工具。你也许会说还有另外一种工具，那就是直觉。但那可能是不理性的，你有可能臆造出任何东西，然后生活在幻觉当中。

So we are asking, very seriously, whether one has understood the nature of thought and whether there is any other action or a way of living which is never fragmentary, never broken up as the world and me, and me and the world. Is there such a state of brain, or a state of non-brain, which is completely holistic, whole?

所以我们在非常认真地问，你是否理解了思想的本质，是否存在另一种行动或者另一种生活方式，它绝不是支离破碎的，绝不会分裂成世界和我、我和世界。是否存在这样一种头脑的状态，或者非头脑的状态，它是彻底完整的、整体的？

You will find out if you are serious, if you are free. You

have to throw away everything that you have accumulated, not physically—please do not throw away your bank account; you won't anyhow—but psychologically put away everything that you have collected. That is going to be very difficult. That means there must be freedom. You know, the word "freedom" etymologically also means love. When there is freedom, boundless and at such enormous depths, there is also love. And to find that out, or to come upon that holistic way of living in which there is no self-interest, there must be freedom from friction, from conflict in relationship.

如果你是认真的，如果你是自由的，你就会发现真相。你必须扔掉你积累起来的一切，不是物质上的——请不要扔掉你的银行账户；你反正也是不会扔掉的——而是从心理上抛开你所积攒的一切。做到这一点会非常困难。这意味着你必须拥有自由。你知道，“自由”这个词在语源上也意味着爱。哪里有自由，广袤无边而又有着巨大深度的自由，哪里就会有爱。而若要找到它，或者邂逅这种没有自私自利的完整的生活方式，就必须从关系中的摩擦和冲突中解脱出来。

We live by relationship. You may live in the Himalayas, or in a monastery, or by yourself in a little banda or a palace, but you cannot live without relationship. Relationship is not just physically or sexually, but to be completely in contact with another. But we are never completely related to another. Even in the most intimate relationship—man and woman—each is pursuing his or her own particular ambition, particular fulfilment, and one's own way of living as opposed to the other, like two parallel lines never meeting. In this relationship there is always conflict. Face the fact.

我们的生活离不开关系。你也许住在喜马拉雅山上或者一所修道院里，或者独自生活在一所小草房或者一座宫殿里，但是你无法脱离关系而生活。关系不只是身体上或者性方面的，而是与另一个人完全地联结在一起，但我们从来没有和别人完全地联结在一起。即使在最亲密的关系中——男人和女人的关系——每个人也都在追逐他或者她自己特别的野心、特别的成就，以及自己与对方对立的生活方式，就像两条永不相交的平行线。这种关系里始终存在着冲突。面对这个事实吧。

And what creates conflict between two human beings? In your relationship with your wife, with your husband, with your children—which is the most intimate relationship—what is it that creates conflict? Ask yourselves, sirs. Is it not that you have an image about your wife and she has an image about you? That image has been built very, very carefully over a short period or a very long period. This constant recording of the brain in relationship with another is the picture that you have created about your wife or your husband. And that picture divides. And especially when you are living in the same house with all the turmoil, you escape from that by becoming a monk or whatever it is. But you have your own problems there too, your own desires, your own pursuits, which again become a conflict.

而又是什么制造了两个人之间的冲突？在你与妻子、与丈夫、与孩子的关系中——这是最为亲密的关系了——是什么制造了冲突？问问你们自己，先生们。难道不是因为你对你妻子抱有意象，而她也对你抱有意象吗？那些意象是在很短或者很长的一段时间内非常非常精心地构建起来的。大脑在与别人的关系中不停进行的这种记录，就是你对自己的妻子或丈夫建立起来的画面，而这些画面造成了分裂。尤其是当你们生活在有着无尽混乱的同一所房子里时，你就会借助成为一名僧侣或者无论什么人来从中逃避。但即使在那里，你还是会有你自己的那些问题，你自己的欲望，你自己的追求，这又再次会成为冲突。

So can you live without a single image of another? No image at all. Have you ever tried it? See the logic of it, the sanity of it that as long as the picture—making machinery goes on, recording the insult or the flattery, it creates an image about another, and that image is a divisive factor. So is it possible to live without a single image? Then you will find out what true relationship is because then there will be no conflict at all in relationship. And that is absolutely necessary if one is to understand the limitation of thought and inquire into a holistic way of living that is completely non-fragmentary.

那么，你能不能在生活中对他人不抱有一丝一毫的意象？完全没有意象。你曾经这样尝试过吗？看看其中的逻辑性、合理性——只要画面制造机械在运转，在记录侮辱或者奉承，就会产生对他人的意象，而这意象是一个造成分裂的因素。所以有可能不带丝毫意象地活着吗？那样你就会发现真正的关系是什么，因为那时关系中就完全不会再有冲突了。

Another factor in our lives is that from childhood we are trained to have problems. When we are sent to school, we have to learn how to write, how to read, and all the rest of it. How to write becomes a problem to the child. Please follow this carefully. Mathematics becomes a problem, history becomes a problem, as does chemistry. So the child is educated, from childhood, to live with problems—the problem of God, problem of a dozen things. So our brains are conditioned, trained, educated to live with problems. From childhood we have done this. What happens when a brain is educated in problems? It can never solve problems; it can only create more problems. When a brain that is trained to have problems, and to live with problems, solves one problem, in the very solution of that problem, it creates more problems. From childhood we are trained, educated to live with problems and, therefore, being centred in problems, we can never solve any problem completely. It is only the free brain that is not conditioned to problems that can solve problems. It is one of our constant burdens to have problems all the time. Therefore our brains are never quiet, free to observe, to look.

我们生活中的另一个因素是我们从小接受的训练就有问题。当我们被送去学校，我们得学习如何读书、如何写字，诸如此类，于是如何写字就变成了孩子的问题。请仔细跟上这些话。数学变成了问题，历史变成了问题，化学也一样。所以孩子从小所受的教育就是和问题生活在一起——神明的问题，一大堆事情的问题。所以说我们的大脑受到的制约、训练和教育，就是和问题生活在一起。我们从小就是这么做的。当大脑是在问题中受到的教育，那会发生什么呢？它永远也解决不了问题，它只会制造更多的问题。当大脑接受的训练就是拥有问题，并且和问题生活在一起，那么当它解决一个问题时，在解决这个问题的过程本身当中，它就会制造出更多的问题。我们从小接受的训练和教育就是和问题生活在一起，所以，既然被问题包围着，我们就永远无法彻底解决任何一个问题。只有不受问题制约的自由的大脑才能解决问题。一直抱有问题，是我们长期背负的重担之一。因此我们的大脑从未安静地、自由地去观察、去看。

So we are asking: is it possible not to have a single problem but to face problems? But to understand those problems, and to totally resolve them, the brain must be free. See the logic of it because logic is necessary, reason is necessary, and only then can you go beyond reason, beyond logic. But if you are not logical, step by step, then you may deceive yourself all along and end up in some kind of illusion. So to find out a way of living in which you can face problems, resolve them, and not be caught in them requires a great deal of observation, attention, and awareness to see that you never deceive yourself for a second.

所以我们在问：有可能没有任何一个问题而是去面对问题吗？然而，若要了解这些问题，并且彻底解决它们，大脑就必须自由。看看这里面的逻辑性，因为逻辑是必要的，理性是必要的，然后你才能超越理性、超越逻辑。然而，如果你不一步步符合逻辑地推理，你也许就会一路欺骗自己，到最后以某种幻觉而告终。所以，若要发现一种生活之道——你能够面对问题、解决问题，而不会被困在其中——就需要广泛地观察、关注和觉知，以确保你绝不会对自己有哪怕一秒钟的欺骗。

First, there must be order. And order begins only when there are no problems, when there is freedom—not freedom to do what you like; that is not freedom at all. To choose between this guru and that guru, or between this book and that book—that is merely another form of confusion. Where there is choice, there is no freedom. And choice exists only when the brain is confused. When the brain is clear, then there is no choice, but only direct perception and right action.

首先你必须拥有秩序。而只有当不存在任何问题，当有自由存在时——不是为所欲为的自由，那根本不是自由——秩序才会出现。在这个古鲁和那个古鲁之间、这本书和那本书之间选择——这只是另一种形式的困惑而已。哪里有选择，哪里就没有自由。而只有当大脑困惑时，选择才会存在。当大脑清晰时，选择就不存在，而是只有直接的洞察和正确的行动。

一九八五年二月二日

学习倾听的艺术

TALK Ⅱ 讲话二

MAY we go on with what we were talking about yesterday evening? This is really not a talk, but a conversation between us, a conversation between two friends-friends who have known each other for a very long time, who are not trying to impress each other or convince each other about anything; we are just friends. We may play golf together, take walks, look at the sky, the trees, the green lawns and the beautiful mountains. And we are talking over our intimate problems-problems which we have not been able to solve, issues that are confusing, living as we do in the modern world, with all its difficulties, turmoil and vulgarity. We are concerned about what human beings are going to become and why, after millions of years, they are what they are now: unreasonable, superstitious, believing in anything, gullible, and caught in organizations.

我们可以继续昨天晚上讨论的内容吗？这真的不是一场演说，而是我们之间的一场对话，两个朋友之间的一场对话——互相认识了很长时间的朋友，谁也不想影响对方或者说服对方任何事情；我们就是朋友而已。我们可以一起打高尔夫球，一起散步，一起凝望天空、树林、绿色的草地和壮丽的群山。而现在我们正一起探讨我们自己的问题——我们从来没能解决的问题，那些令人困惑的事情，我们生活在这个有着无尽的困苦、混乱和粗俗的现代社会中遇到的那些问题。我们关心的是人类将会成为什么，为什么在过了数百万年之后他们会变成现在的样子：不理性、迷信、相信一切、容易上当受骗，并且困在各种组织之中。

So you and the speaker are going to talk over things together. That is, you do not merely listen to the speaker, but enter into the spirit, into the inquiry. So you have to exercise your brains as much as possible. Do not accept anything he says. Be sceptical, question, inquire and, if you will, together we shall take a long journey, not only outwardly but inwardly, into the whole psychological world: the world of thought, the world of sorrow, the world of fear and travail. This is not a lecture to inform or instruct you. But together we are going to have a dialogue, a conversation, without holding on to our own particular beliefs or convictions, experiences or superstitions, but exchanging, changing as we go along. So there is no question of doing any propaganda, of trying to convince you of anything at all. On the contrary, we must doubt, question, inquire and, as friends, listen to each other.

所以说你和讲话者将一起探讨一些事情。也就是说，你不只是单纯地在听讲话者说话，而是要进入探询之中，进入这些话的精神之中。所以你必须尽可能充分地运用你的大脑。不要接受他说的任何事情，保持怀疑，去质疑、去探询，如果你愿意，我们将一起踏上一段长长的旅程，不仅从外在，而且从内在，深入探究整个心理世界：思想的世界，悲伤的世界，恐惧和辛劳的世界。这不是一场要指导你或者向你灌输知识的讲座，而是我们要一起开展一场对话、一次交谈，不要紧紧抱持着你自己特定的信仰或信念、经验或迷信，而是当我们一起前行时，我们互相交流，同时也在改变。所以这里不存在进行任何宣传的问题，也完全不想说服你相信任何事情。正相反，我们必须质疑、怀疑、探询，并且像朋友一样倾听彼此。

Listening is an art which very few of us are capable of. We never actually listen. The word has a sound and when we do not listen to the sound, we interpret it, try to translate it into our own particular language or tradition. We never listen acutely, without any distortion. So, the speaker suggests, respectfully, that you so listen and not interpret what he says. When you tell a rather exciting story to a little boy, he listens with a tremendous sense of curiosity and energy. He wants to know what is going to happen, and he waits excitedly to the very end. But we grown-up people have lost all that curiosity, the energy to find out, that energy which is required to see very clearly things as they are, without any distortion. We never listen to each other.

You never listen to your wife, do you? You know her much too well, or she you. There is no sense of deep appreciation, friendship, amity, which would make you listen to each other, whether you like it or not. But if you do listen so completely, that very act of listening is a great miracle.

倾听是一门我们很少有人能够掌握的艺术，我们从不真正地倾听。词语有自己的声音，而当我们不去倾听那个声音时，我们就会诠释它，试图把它翻译成我们自己特有的语言或者传统。我们从不毫无扭曲、敏锐地倾听。所以讲话者充满敬意地建议你能够这样倾听，而不诠释他说的话。当你给一个小男孩讲一个非常有趣的故事时，他会带一股极大的好奇心和活力去听。他想知道接下来会发生什么，他会一直兴奋地等到最后。但我们成年人已经完全失去了那份发现真相的好奇心和活力，那股能非常清晰地、如实地、毫无扭曲地看到事物所需要的能量。我们从不倾听彼此。你从不倾听你的妻子，对吗？你对她或者她对你都太熟悉了。你们之间没有那股深深的理解、友谊、和睦，这些品质足以让你们倾听彼此，无论你是否喜欢。然而如果你确实能够如此完全地倾听，那倾听的行为本身就是一个伟大的奇迹。

That listening, like seeing, observing, is very important. We never observe. We observe things that are convenient, friendly. We observe only if there is a reward or punishment. I do not know if you have noticed that our whole upbringing, all our education and our daily life is based on one principle: reward and punishment. We meditate in order to be rewarded, we"progress"in order to be rewarded, and so on. When we seek a reward, physical or psychological, in that search for a reward there is also the punishment—if that reward is not satisfying. So could we listen to each other perse, for itself, not for something else? Could we listen, as we would listen to marvellous music or to the song of a bird, with our hearts, with our minds, with all the energy that we have? Then we can go very far.

这份倾听，就像看和观察一样，是非常重要的。我们从不观察。我们只观察方便的、友好的事物。只有存在奖惩的时候我们才去观察。我不知道你有没有注意到，我们的整个成长环境，我们所有的教育和我们的日常生活都基于同一个原则：奖励和惩罚。我们冥想是为了得到奖赏，我们进步是为了得到奖赏，等等。当我们寻求奖赏时，无论是身体上的或者心理上的，而在这份对奖赏的追求中同样存在着惩罚——如果这份奖赏不令人满意的话。所以我们能不能为了倾听本身去倾听彼此，而不是为了别的东西？我们能不能用我们的心、我们的头脑、我们的所有能量去倾听，就像我们聆听美妙的音乐或者聆听小鸟的歌唱一样？那样我们才能走得很远。

Most human beings, all of us, seek security, and it takes many forms. Security is very important. If we are not secure, both physically and psychologically, our brains cannot function adequately, fully, energetically. We must have security. But physical security is denied to millions and millions of people; they have hardly one meal a day. And we, the so-called educated, well-to-do people are all the time seeking, consciously or otherwise, a kind of security which would give us complete satisfaction. We want security, and it is necessary both biologically and psychologically. But in our search for security we never inquire into what is insecurity. If we can find out together what insecurity is and why we are insecure, then in its very unfolding, security comes about naturally.

大多数人，我们所有人，都寻求安全，而它有很多种形式。安全非常重要。如果我们身体上和心理上都不安全，我们的大脑就无法恰当地、充分地、有活力地运转。我们必须拥有安全。但是有千百万人被剥夺了身体上的安全，他们一日难得一餐。而我们，所谓受过教育的、生活富裕的人们，也一直在有意识或无意识地追求一种能够带给我们彻底满足的安全。我们希望得到安全，它在生理上和心理上都是必不可少的。但是我们在追求安全的过程中却从不探究不安全是什么。如果我们能够一起弄清楚不安全是什么，以及我们为什么感到不安全，那么在它展现出来的过程中，安全就自然而然地到来了。

So what is insecurity? Why are we insecure in our relationship to each other? There is tremendous disturbance, turmoil and agony in the external world, and each one wants his own place, his own security, and wants to escape from this terrible state of insecurity. So, can we, together inquire into why we are insecure? Not into what security is because your security may be an illusion. Your security may be in some romantic concept, in some image, tradition, or in a family and name. What does that word"insecure"mean? In your relationship to your wife or husband, there is not a sense of complete security. There is always this background, this feeling that everything is not quite right. So inquire with me into why human beings are insecure. Is it about not having a job? In a country like this, which is overpopulated, there are probably ten thousand people for one job. Don't you know all this, or am I inventing it? If we were not insecure, we would not talk about gods, we would not talk about security. Because we are insecure, we seek the opposite.

那么，不安全是什么？在相互关系中我们为什么感觉不安全？外部世界中有着可怕的不安、混乱和痛苦，每个人都想要他自己的空间、他自己的安全，想要逃离这可怕的不安全状态。所以，我们能否一起来探询我们为什么感觉不安全？——而不是探究安全是什么，因为你的安全感或许是个幻觉。你的安全或许存在于某个浪漫的理念、某个意象和传统之中，或者家庭和头衔之中。“不安全”这个词的含义是什么？在你与妻子或丈夫的关系中，并没有彻底的安全感。总是有这种背景，这种感觉，觉得一切都好像不太对劲。所以，请和我一起来探询人类为什么不安全。是因为没有工作吗？在这样的一个国家里，人口过剩，也许有一万人在争夺一份工作。这一切你们难道不知道吗？还是都是我虚构出来的？如果我们并非不安全，我们就不会谈论神明，我们就不会谈论安全。因为我们不安全，所以我们追求它的反面。

Have you ever listened to sound? Sound! The universe is

filled with sound. The earth is full of sound. And we seek silence. We meditate to find some kind of peace or some kind of silence. But if we understood sound, in the very hearing of the sound there would be silence. Silence is not separate from sound. But we do not understand that because we never listen to sound. Have you ever sat under a tree when the air is very still, quiet, when not a leaf is dancing? Have you ever sat under a tree like that and listened to the sound of the tree? If there was no silence, there would be no sound. So the sound of insecurity—the very sound—makes us seek security because we have never listened to the sound of insecurity. If we listened to all the implications of insecurity, to the whole movement of insecurity—which makes us invent gods, rituals and all that—then out of that insecurity there comes about, naturally, security. But if you pursue security as something separate from insecurity, then you are in conflict.

你可曾聆听过声音？声音！宇宙中充满了声音，大地充满了声音，而我们却追寻寂静。我们冥想，以期盼找到某种安宁或者某种寂静。然而如果我们懂得了声音，在对声音本身的聆听之中就会有寂静存在。寂静与声音并不是分开的。但我们并不懂得这一点，因为我们从未聆听声音。当空气中非常安宁、非常寂静，没有一片树叶在飞舞时，你可曾就这样坐在一棵树下，倾听树的声音？如果没有寂静，也就不会有声音。所以，是不安全的声音——这声音本身——让我们去寻找安全的，因为我们从未聆听过不安全的声音。如果我们倾听了不安全的所有含义，倾听了不安全的整个运动——是它让我们发明了神明、仪式等这一切——那么从那不安全之中，安全就会自然而然地到来。但是，如果你把安全当作某种与不安全相分离的东西来追求，你就陷入了冲突。

You know, of an evening when the sky is clear—not in

Bombay—and there is only one star in the sky, and there is absolute silence, if you listen to that silence, in that silence there is sound. And there is no separation between sound and silence; they both go together. In the same way, understand insecurity, its causation. The cause of insecurity is our own limited, broken-up psychological state. But when there is a way of living that is holistic, then there is no such thing as security or insecurity.

你知道，天空晴朗的夜晚——不是在孟买——天上只有一颗星星，此时有一种绝对的寂静，如果你聆听那寂静，那寂静之中是有声音的。声音和寂静之间没有分隔，它们是并肩而行的。以同样的方式去理解不安全以及它的原因。不安全的原因就是我们自己局限的、破碎的心理状态。然而，当存在一种完整的生活方式，那么诸如安全或不安全之类的事情就会不复存在。

So, if you will, we shall talk over together: What is a holistic way of life? The word whole means complete, a state in which there is no fragmentation—no fragments such as a businessman, an artist, a poet, a religious person, and so on. But we are constantly categorizing people as communists, socialists, capitalists, and so on.

所以，如果你愿意，我们就一起来探讨：什么是完整的生活方式？“完整”这个词意味着“完全”，意味着一种没有破碎的状态——没有诸如生意人、艺术家、诗人、宗教人士等之类的碎片。可是我们不停地把人们归类为共产主义者、社会主义者、资本主义者等。

Our lives, if you observe closely, are broken up. Our lives are fragmented. And we have to understand why we human beings, who have lived on this marvellous earth for millions of years, are so fragmented, so broken up. As we said yesterday, one of the main causes of this breaking up is that the brain is a slave to thought, and thought is limited. Wherever there is limitation, there must be fragmentation. When I am concerned with myself, with my progress, my fulfilment, my happiness, my problems, I have broken up the whole structure of humanity into the“me”. So one of the factors of why human beings are fragmented is thought. And another of the factors is time.

如果我们仔细观察的话，会发现我们的生活是分裂的，我们的生活是支离破碎的。而我们必须弄明白为什么我们人类在这个神奇的地球上生活了数百万年，却依然如此支离破碎、如此分崩离析。正如我们昨天所说，这种分裂的一个主要原因就是大脑是思想的奴隶，而思想是有限的。当我只关心我自己，只关心自己的进步、自己的成就、自己的快乐和自己的问题，我就已经把人类的整个结构打破了，变成了“我”。所以人类为什么分崩离析的一个因素就是思想。而另一个因素就是时间。

Have you ever considered what time is? According to the scientists who are concerned with it, time is a series of movements. So movement is time. And time is not only by the watch, chronological: time as the sun rising and the sun setting, the darkness of the night and the brightness of the morning. There is also psychological time, inward time:"I am this, but I will become that"; "I don't know mathematics, but one day I will learn all about it."That requires time. To learn a new language requires a great deal of time. There is time to learn, to memorize, to develop a skill, and there is also time as the self-centred entity saying, "I will become something else." The"becoming"psychologically also implies time. We are inquiring into not only the time to learn a skill but also the time which we have developed as a process of achievement. You do not know how to meditate, so you sit cross-legged and learn how to control your thoughts so that one day you will achieve what meditation is supposed to be. So you practise, practise, practise, and thus you become mechanical. That is, whatever you practise makes you mechanical.

你可曾思考过时间是什么？根据有关科学家的说法，时间是一系列的运动，所以运动就是时间。而时间不仅仅是钟表上、时序上的时间：日升日落的时间，黑暗的夜晚和明亮的清晨这样的时间，此外还有心理上的时间、内在的时间："我是这样的，但我要变成那样"；"我不懂数学，但有一天我会弄懂的。"这都需要时间。学一门新语言需要大量的时间。存在着学习、记忆、锻炼技能所需要的时间，还有自我中心的存在体说"我要成为别的什么"这样的时间。这种心理上的"成为"

也隐含了时间。我们探究的不仅仅是学会一门技能需要的时间，还有我们作为成就的过程建立起来的时间。你不知道如何冥想，于是你盘腿坐下，学习如何控制自己的思想，以期待有一天你会达到事先设想的那种冥想。所以你练习、练习、练习，于是你变得机械化，也就是说，无论你练习什么，都会让你变得机械化。

So time is the past, the present, and the future. Time that is the past is all the memories, all the experiences, knowledge, and all that human beings have achieved. All that which remains in the brain as memory is the past. That is simple. The past—the memories, the knowledge, the experiences, the tendencies—the background is operating now. So you are the past. And the future is what you are now, perhaps modified. The future is the past modified. See this, please understand. The past, modified in the present, is the future. So if there is no radical change in the present, tomorrow will be the same as what you are today. So the future is now—not the future needed for acquiring knowledge, but the psychological future. That is, the psyche, the"me", the self, is the past, is memory. That memory modifies itself now, and goes on. So the future and the past are in the present. All time—the past, the present and the future—is contained in the now. This is not complicated please, it is logical. So if you do not change now, instantly, the future will be what you are now, what you have been. So is it possible to change radically, fundamentally, now? Not in the future.

所以说时间就是过去、现在和未来。过去的时间就是所有的记忆、所有的经验和知识，以及人类所得到的一切。大脑里作为记忆留存下来的一切，就是过去，这很简单。过去——记忆、知识、经验、倾向——这个背景现在正运转着，所以你就是过去。而未来就是你现在的样子，也许会稍作调整，未来就是调整之后的过去。请看到这一点，理解这一点。过去在此刻得到调整之后，就是未来。所以，如果现在你没有发生根本的转变，明天你还会跟今天一个样。所以未来就是现在——不是获取知识所需要的那种未来，而是心理上的未来。也就是说，心智、“我”、自我，就是过去，就是记忆。这记忆在此刻调整自己，然后再延续下去。所以未来和过去都在此刻。所有的时间——过去、现在和未来——都包含在此刻之中。拜托，这并不复杂，这是合乎逻辑的。所以，如果你现在没有发生即刻的转变，那么未来就会是你现在的样子、你过去的样子。那么，现在——而不是未来——有可能发生彻底的、根本的转变吗？

We are the past. There is no question about it. And that past gets modified by reaction, by challenges, in various ways. And that becomes the future. Look, you have had a civilization in this country for three to five thousand years. That is the past. But modern circumstances demand that you break away from the past, and so you have no culture any more. You may talk about your past culture and enjoy its past fame, but that past is blown up, scattered by the present demands, the present challenge. And that challenge, that demand is changing it into an economic entity.

我们就是过去，这一点毫无疑问。而这个过去通过各种方式，借助反应和挑战得以调整，然后就变成了未来。你瞧，这个国家拥有三千到五千年的文明，这就是过去。而当代的环境要求你脱离过去，于是你们就再也没有了文化。你们也许会谈论过去的文化，对它过去的名声津津乐道，但那过去已经被当今的需求、当今的挑战摧毁了、拆散了。而这种挑战、这种需求正把它转变成一个经济实体。

So all past and future is in the now. So all time is in the now. And we are saying that thought and time are the major causes of fragmented human beings. Also, we want roots, identification. We want to be identified with a group, a guru, a family, a nation, and so on. And the threat of war is a major factor in our lives. War may destroy our psychological roots, and therefore we are willing to kill others. So these are the major factors of our fragmented lives.

所以说所有的过去和未来都在此刻，所有的时间都在此刻。而我们说思想和时间是人类破碎不堪的主要肇因。同时，我们也想要根基和认同，我们希望与一个团体、一名古鲁、一个家庭、一个国家等相认同。而战争的威胁也是我们生活中的一个主要因素。战争也许会摧毁我们心理上的根基，因此我们愿意去杀害别人。所以说这些都是我们支离破碎的生活的主要因素。

Now, do you listen to the truth of it, or merely to a description of what is being said and, therefore, carry the description and not the truth, the idea and not the fact? For instance, the speaker says, "All time is now."If you understand that, it is a most marvellous truth. Do you listen to that as a series of words, as an idea, as an abstraction of the truth, or do you capture the truth of it? Which is it that you are doing? Do you see, live with the fact? Or do you make an abstraction of the fact, an idea, and then pursue the idea and not the fact? That is what the intellect does. Intellect is necessary, but probably we have very little intellect anyhow because we have given ourselves over to somebody. So when you hear a statement like,"All time is now", or"You are the entire humanity because your consciousness is one with all the others", how do you listen to it? Do you make an abstraction, an idea, of it? Or do you listen to the truth, to the fact of it, to the depth of it, to the sense of immensity involved in it? Ideas are not immense, but a fact has got tremendous possibility.

那么，你有没有听到这些话中包含的真相，还是说你仅仅听到了这些描述的话，因而带走的是描述而不是真相、理念或者事实？比如，讲话者说：“所有的时间都是现在。”如果你理解了这一点，那它就是一个非凡无比的真理。你是把这句话当作一系列的词语，当作一个观念，当作对真理的抽象表达来听的呢，还是说你捕捉到了其中的真理？你所做的是哪一样？你是否看到了事实，并与事实共处？还是说，你从事实中得到了一个抽象的概念，然后再去追逐那个概念而不是事实本身？这就是智力所做的事情。智力是必要的，但也许我们就是智力很低下，因为我们把自己交付给了别人。所以，当你听到像“所有时间都是现在”，或者“你就是全体人类，因为你和其他所有人共有同一个意识”这样的说法，你是怎么听的呢？你是从中得出了一个抽象的概念或者想法吗？还是说，你聆听了其中的真理、其中的事实、其中的深度，聆听了其中所包含的那种无限？观念并不是无限的，而事实却拥有无限的可能性。

So a holistic life is not possible when there is thought, time, and the desire for identification and for roots. They prevent a way of living that is whole, complete. You hear the statement, and your question then will be, "How shall I stop thinking?" It is a natural question, isn't it? You know that time is necessary to learn a skill, a language, or a technical subject. But you have also just begun to realize that the "becoming", the moving from "what is" to "what should be", involves time, and that it may be totally wrong, it may not be true. So you begin to question. Or do you just say, "I don't understand what you are talking about but I will go along with it" ? Which is actually what is taking place. Honesty, like humility, is one of the most important things. When a vain man cultivates humility, that humility is part of vanity. But humility has nothing to do with vanity, with pride. It is a state of the mind that says, “I don’t know what I am, let me inquire”, and never says, “I know”.

所以，当思想、时间以及对认同和根基的渴望依然存在时，完整的生活就不可能实现。它们妨碍了完整的、整体的生活方式出现。你听到了这个说法，然后你就会问：“我要如何停止思考呢？”这是一个很自然的问题，不是吗？你知道时间对于学习一门技能、一门语言或者一个技术上的学科是必要的。但你也开始意识到“成为”，这种从“现在如何”到“应该如何”的运动中也包含了时间，而这种时间也许是完全错误的，也许是不真实的。于是你开始质疑。还是你只是说，“我不明白你在讲什么，但我会带走这个说法”？这就是实际上正在发生的事情。诚实，就像谦卑一样，是最重要的东西之一。当一个虚荣的人培养谦卑，那谦卑就是虚荣的一部分。但谦卑实际上与虚荣，与骄傲毫无关系。那是这样的一种心灵状态：它会说“我不知道自己现在如何，让我来探究一下”，而绝不会说“我知道了”。

Now, you have listened to the fact that all time is now. You may agree or you may not agree. That is a dreadful thing—agreeing and disagreeing. Why should we agree or disagree? The fact is that the sun rises in the east; you don't agree or disagree with it. So can we put aside our conditioning of agreeing and disagreeing so that we both can look at facts, so that there is no division between those who agree, and those who do not? Then there is only seeing things as they are. You may say, "I don't see", but that is a different matter. Then we can go into why you don't see. But when we enter into the area of agreement and disagreement, we become more and more confused. The speaker has said our lives are fragmented, our ways of thinking are fragmented. You are a businessman, you earn lots and lots of money and then you build a temple or give to charity. See the contradiction in it. We are never deeply honest with ourselves—not honest in order to be something else or to understand something else. But to be clear and to have an absolute sense of honesty, which is to have no illusions. If you tell a lie, you tell a lie and you know it and say,"I have told a lie", and do not cover it up. When you are angry, you are angry and you say that you are angry. You do not find causes, explanations for it, or try to get rid of it. This is absolutely necessary if you are going to inquire into much deeper things, as we are doing now. Not make a fact into an idea but to remain with the fact—that requires very clear perception.

现在，你已经听到了这个事实：所有的时间都是现在。你也许会同意，也许会不同意。这真是一件致命的事情——同意和不同意。我们为什么要同意或者不同意呢？太阳会从东方升起，这是一个事实，对此你不需要同意或者不同意。所以，我们能不能抛开我们同意或是不同意的制约，这样我们双方就都能看着事实，于是同意和不同意的人之间就没有了分裂？这样的话就只剩下如实地看到事情本身。你可以说，“我没看到”，但那就是另外一回事了。然后我们就可以来看看你为什么没看到。然而，当我们进入了同意不同意的领域，我们的思维方式就会变得支离破碎。你是一个生意人，你赚了很多很多钱，然后你建了一座寺庙或者捐献给慈善事业。看看其中的矛盾。我们从来没有对自己真正诚实过——不是为了成为别的什么或者领悟别的什么才诚实，而是保持清晰，拥有一种绝对的诚实感，那就是没有任何幻觉。如果你说了谎，那你就是说了谎，你知道这一点，然后你说“我说谎了”，而不把它掩盖起来。当你生气时，你就是生气了，然后你说你生气了。你不去为它寻找借口、寻找解释，也不试图消除它。如果你想探究更为深刻的事情，就像我们现在所做的这样，那么这一点就是绝对必要的。不要把事实变成概念，而是与事实待在一起——而这需要非常清晰的洞察。

Now, having heard all this you will say:“Yes, I understand this logically, intellectually.” And you will ask:“How am I to relate what I have understood logically, intellectually, to what I have heard? What is the truth?” So you have already created a division between intellectual understanding and action. Do you see this? So listen, just listen. Don’t do anything about it. Don’t ask, “How am I to get something?”“How am I to put an end to thought and time? ”—which you cannot. That would be absurd because you are the result of time and thought. You will just go round and round in circles. But listen, don't react, don't ask, "how?"but actually listen—as you would listen to some lovely music or to the call of a bird—to the statement that all time is in the now and that thought is a movement. Thought and time are always together. They are not two separate movements, but one constant movement. That is a fact. Listen to it.

现在你听到了这一切，你会说："是的，我从逻辑上、智力上理解了这一点。"然后你会问："我要如何把我从逻辑上、智力上理解的东西，和我听到的东西联系到一起呢？真理是什么？"所以你已经在智性的理解和行动之间制造了一种分裂。你看到这一点了吗？所以请听我说，只是倾听，不要对我说的话做任何事情。不要问，"我要如何才能得到些什么？""我要如何结束思想和时间？"——你做不到。这个问题很荒唐，因为你就是思想和时间的结果。那样你只会来来回回地兜圈子。但是倾听，不作反应，也不问“如何”，而只是实实在在地倾听——就像你聆听一段美妙的音乐或者鸟儿的鸣叫那样——来倾听“所有时间都在此刻”和“思想是一种运动”这些说法。思想和时间总是并肩而行，它们并非各自独立的两种运动，而是一种持续不断的运动。这是一个事实，请听一听。

Then, you want to be identified, and that is one of the causes of fragmentation of our lives, like time and thought. Also, you want security and therefore you take roots. So these are the factors of fragmentation. Listen to this. Don’t do anything. Now, if you listen very carefully, that very listening creates its own energy. If you listen to the fact of what is being said and do not react—because you are just listening to it—that implies the gathering of all your energy to listen. That means giving tremendous attention to listening. And that very listening breaks down the factors or the causes of fragmentation. If you do something, then you are acting upon the fact. But if there is an observation, without distortion, without prejudice, then that very observation, that very perception which is great attention, burns away the sense of time, thought, and all the rest of it.

然后你还希望得到认同，这也是我们的生活支离破碎的原因之一，就像时间和思想一样。此外，你还希望得到安全，进而扎下根来。所以这些都是破碎的因素。请听一听这些，不要做任何事情。那么，如果你非常仔细地倾听，那倾听本身就会创造出它自身的能量。如果你倾听现在所说的事实并且不做反应——因为你只是在单纯地倾听——那就意味着集中你所有的精力去倾听，意味着为倾听灌注了非凡的注意力，那么这倾听本身就摧毁了破碎的各个因素或者根源。如果你做了什么，那么你就是在对事实采取行动。然而，如果存在没有扭曲、没有偏见的观察，那么本身即为巨大关注的那观察就烧光了时间感、思想，以及诸如此类的一切。

And also one of the factors of our lives, in which we live in fragmentation, is fear. That is a common human fact. Human beings, right from the beginning of time, have been frightened. And they have never solved the problem. If you were not frightened at all, there would be no gods, no rituals, no prayers. It is our fear that has created all the gods, all the deities, and the gurus with their absurdities. So can we go into the question of why human beings live in fear and whether it is possible to be free of it entirely, not occasionally, not sporadically? Can you be aware of the objects of fear and also the inward causes of fear? You may say,“I am not afraid”, but all your background is structured on fear.

造成我们生活在支离破碎中的另一个因素，就是恐惧。恐惧是人类共有的一个事实，人类从存在伊始就已经心存恐惧了，但他们从未解决这个问题。如果你根本不恐惧，就不会有各种神明、仪式和祈祷。正是我们的恐惧制造了所有的神明和所有的古鲁，以及它们荒唐的一切。所以，我们能不能探究一下这些问题：为什么人类生活在恐惧中？有没有可能彻底而不是有时候、偶尔地摆脱恐惧？你能不能觉察到恐惧的对象以及恐惧的内在根源？你也许会说，“我不害怕”，但你的所有背景都构筑在恐惧之上。

What is fear? Are you not afraid? If you are really honest, for a change, will you not say,“I am afraid”? Afraid of death. Afraid of losing a job. Afraid of your wife or husband. Afraid of public opinion. Afraid of not being recognized by your guru as a great disciple. Afraid of the dark. Afraid of so many things. We are not talking about the objects of fear, fear of something. We are inquiring into fear perse, in itself. So we are asking: What is the cause of fear, and what is fear without a cause? Is there such a thing as fear without a cause? Or does the word"fear", the sound of fear, evokes in us fear? For example, when you hear the word"communism", you would react to it, if you are a capitalist or even if you are a socialist. And when you hear the word"fear", you react to it, don't you? Of course you do.

恐惧是什么？你难道不害怕吗？我们换个说法，如果你真正诚实的话，你难道不会说“我害怕”吗？害怕死亡，害怕失业，害怕你的妻子或丈夫，害怕公众舆论，害怕没有被你的古鲁认可为一名出色的弟子，害怕黑暗，害怕如此之多的事情。我们谈的不是恐惧的对象，不是对什么的恐惧，我们探究的是恐惧本身。所以我们问：恐惧的根源是什么？没有原因的恐惧又是什么？存在没有原因的恐惧这回事吗？还是说，“恐惧”这个词，恐惧的声音，在我们内心唤起了恐惧？比如说，当你听到“共产主义”这个词，无论你是一个资本主义者，还是一个社会主义者，你都会对它做出反应。而当你听到“恐惧”这个词，你也会对它做出反应，不是吗？你当然会。

Now, does the word create the fear, or is the word different from the fear? We are asking, is the word"fear"different from the fact, or does the word create the fact? One must be clear about this. If there was no such word as"fear", would there be fear? You see, sir, the word"love"is not that flame. Similarly, the word"fear"may not be the actual, the sense of being gripped, of living in a state of nervousness. You know what fear does to people. They live in darkness, they are all the time frightened, frightened, frightened, and their lives get shattered. So we are saying that the word is not the fact, the word is not the thing. That must be quite clear. So what is the cause of fear? Now, just a minute. The speaker has asked you this question: What is the cause of fear? How do you listen to it? That question has a vitality of its own, an energy of its own. It is a very serious question, not merely an intellectual one. If you remain with the question and not try to find an answer, the question itself begins to unfold.

那么，是这个词制造了恐惧，还是这个词和恐惧是不同的？我们问的是，“恐惧”这个词与这个事实是不同的吗？抑或是这个词造成了这个事实？你必须清楚这一点。如果没有“恐惧”这个词，还会有恐惧吗？你瞧，先生，“爱”这个词并不是那爱的火焰。同样，“恐惧”这个词也许并不是事实，并不是那种被攫住、处于紧张状态中的感觉。你知道恐惧对人们做了什么。他们活在黑暗中，他们始终战战兢兢、小心翼翼，他们的生活被击垮了。所以我们说词语并不是事实本身，那个词并不是那件事。这一点必须非常清楚。那么恐惧的根源是什么？现在请等一下。讲话者刚刚问了你这个问题：恐惧的根源是什么？你是如何倾听这个问题的？这个问题有它自己的生命力、它自己的能量。这是一个非常严肃的问题，而不只是一个智力问题。如果你与这个问题待在一起而不试图找到答案，那么问题自己就会开始展现出来。

Suppose I tell you in all seriousness,"I love you". I say that with my heart. How do you listen to it? Do you listen to it, or do all your reactions come into it? Perhaps, you have never loved at all. You may be married, have sex, and have children, but you may not know what love is. Probably you don't. This may be a fact. If you loved, there would be no images, no divisions. So what is the cause of fear? If I may most respectfully suggest, listen to it. Put that question to yourself and do not try to find an answer. Because if you try to find an answer—which is to find out the cause and then end it—it means that“you”are different from fear. But are you different from fear? Or you are fear? If you are greedy, is the greed different from you? When you are angry, is anger different from you? You are anger, you are greed. So you are fear. Of course. Can you admit—not admit but see the fact—that anger is you, greed is you, fear is you? But now you have separated yourself from fear and so you say,“I must do something about it”. And you have done something about it for fifty thousand years; you have invented gods, puja, and all the rest of it. So listen to the question and don’t react. Don’t ask“how?”The word“how?”must disappear completely from your minds. Otherwise you would be asking for help. Then you become dependent on somebody. Then you lose all your vitality, independence, and sense of stability.

假设我极其认真地告诉你，“我爱你”，我用心地说出这句话，你会如何去听呢？你会倾听它吗，还是说你所有的反应都介入了进来？也许你从来都没有爱过。你也许结了婚，有性生活，有孩子，但你也许并不知道爱是什么。也许你不知道，也许这就是事实。如果你爱过，那就不会有意象，不会有分裂。那么恐惧的根源是什么？如果我可以满怀敬意地建议，请听一听这个问题。向自己提出这个问题，并且不去寻找答案。因为如果你试图找到答案——也就是找到根源然后终结它——那就意味着“你”和恐惧是不同的。然而你和恐惧是不同的吗？还是说你就是恐惧？如果你贪婪，那贪婪与你不同吗？当你愤怒时，愤怒与你不同吗？你就是愤怒，你就是贪婪，所以你就是恐惧。当然如此。你能不能承认——不是承认，而是看到事实——愤怒就是你，贪婪就是你，恐惧就是你？但现在你把自己和恐惧分离了开来，所以你说，“我必须对它做点儿什么”。而你已经对它做了五万年了，你发明了各种神明、仪式以及诸如此类的一切。所以请倾听问题而不要做出反应，不要问“如何”。“如何”这个词必须从你头脑中彻底消失，否则你还会寻求帮助，接下来你就会依赖别人，然后你就失去了你所有的生命力、独立性以及稳定感。

So will you put this question to yourself and not expect an answer? Put the question. You have planted a seed in the earth, and if the seed is alive it will go through concrete. Haven’t you seen a blade of grass in the pavement? What extraordinary vitality that blade of grass has to break through the heavy cement! In the same way, if you put this question to yourself and hold it, then you will see the cause. The cause is very simple. I can explain it, but that is not the point for the moment. What is important is to put the question because you are serious and you want to find out. Let the question itself answer—like the seed in the earth. Then you will see that the seed flowers and withers. Don’t pull it out all the time to see if it is growing. As you have planted a seed in the earth, so we have planted in our hearts and minds the sense of what is fear. But if you keep on pulling at it, and asking it, then you lose energy. But if you leave the question alone and live with it, then you will see that there is a cause for fear—not the word, not the explanation but the actual truth of it. The cause of fear is thought and time:"I have a job but I might lose it tomorrow.""I have lived with pain and it is gone now, but I am afraid it might come back."Don't you know all this? So time is the future and the past, as I explained now. And also thought.

所以，你愿不愿意向自己提出这个问题，并且不期待得到答案？提出这个问题，你就已经把种子种到泥土里了，而如果那颗种子有生命力，它就会钻破水泥。你难道没有看到过人行道上钻出的青草叶片吗？那片草叶具有怎样非凡的生命力才能穿破厚重的水泥啊！

同样，如果你向自己提出这个问题并紧握着它，那么你就会看到恐惧的根源。那根源非常简单，我可以解释它，但重点暂时不在这里。重要的是提出问题，因为你很认真，你想弄清真相，让问题自己来回答——就像泥土里的种子一样。然后你就会发现那颗种子成长绽放然后枯萎了。不要老是把它拔出来看看它是不是在生长。就像你把种子种进了泥土，我们也已经在自己的内心和头脑中种下了“恐惧是什么”这个问题的含义。但是如果你老是把它拔出来并且追问，那么你就失去了活力。然而，如果你把问题留在那里并与它共处，你就会发现恐惧的根源出现了——不是词语，不是解释，而是它实实在在的真相出现了。恐惧的根源就是思想和时间：“我有一份工作，但明天我可能会失业。”“我遭受过痛苦，现在痛苦消失了，可我害怕它还会回来。”你难道不知道这些吗？所以说时间就是未来和过去，就像我现在解释的这样。另外还有思想。

Thought and time are the two factors of fear. You cannot do anything about it. Don't ask,"How am I to stop thinking?"It is too silly a question. Because you have got to think—to go from here to your house, to drive a car, to speak a language. But time may not be necessary at all psychologically, inwardly. So we are saying fear exists because of the two major factors of time and thought, in which is involved reward and punishment. Now, I have heard this statement made by you. And I have listened to it so immensely because it is a tremendous problem which man has not solved at all and which, therefore, is creating havoc in the world. I have listened to you, listened to the statement. And you have also told me: Don't do anything about it; just put forward the question and live with it, as a woman bears the seed in her womb. So you have put forward the question. Let that question flower. In the flowering of that question, there is also the withering away of that question. It is not the flowering and then the ending—the very flowering is the ending. Do you understand what we are talking about?

思想和时间是恐惧的两个要素。你对它们什么也不能做。不要问："我该如何停止思考？"这是个太过愚蠢的问题，因为你必须得思考——从这里走到你的房子，开车，说一门语言，都需要思考。但是从心理上、从内在来说，也许根本不需要时间。所以我们说，恐惧之所以存在，是因为两个主要的因素——时间和思想，其中涉及了奖赏和惩罚。现在，我听到了你说的这句话。我如此全神贯注地倾听了你说的话，因为恐惧是人类从未解决的一个巨大问题，并因此在世界上制造了可怕的破坏。我倾听了你，听到了那句话。同时你又告诉我：对它什么也不要做；只是提出这个问题并与之共处，就像一个女人的子宫里孕育着那颗种子一样。于是你提出了那个问题，让它自己绽放。在绽放之中，那个问题同时也在枯萎。不是先有绽放，然后再终结——绽放本身就是终结。你明白我们在说什么吗？

Sirs, learn the art of listening, to your wife, to your husband. Listen to the man in the street—his hunger, his poverty, his desperation, and the lack of love. Listen to it. When you listen, at that moment you have no problems, you have no turmoil. You are just listening and, therefore, there is no time in the act of listening.

先生们，请学习倾听的艺术，倾听你的妻子、你的丈夫，倾听街上的那个人——他的饥饿，他的贫穷，他的绝望，以及爱的匮乏。去倾听。当你倾听时，在那一刻你没有问题，没有混乱。你只是在倾听，因此倾听的行动中没有时间存在。

一九八五年二月三日

爱、悲伤和死亡 TALK Ⅲ 讲话三

WE were talking last week about a holistic life-a way of living that is not fragmented, not broken up, as our lives are. We went into the question fairly deeply. We saw what the causes of this fragmentation were, the various factors in our social, moral and religious life that have broken us up as Hindus, Muslims, Christians, Buddhists, and so on. Religions have been greatly responsible for this catastrophe. We also talked about time, time being the past-all the memories, the accumulated experiences, and so on-modified by the present and continuing into the future. That is our life. We have existed on this earth, according to the biologists and the archaeologists, for million years. During that long period of so-called evolution, we have accumulated enormous memory. We also talked about the limitation of memory and thought, and about how that limitation has broken up the world geographically, nationally, and religiously.

我们上周谈到了完整的生活—— 一种不破碎、不分裂的生活方式，完全不像我们如今这样的生活。我们非常深入地探讨了这个问题。我们看到了这种破碎的根源是什么，我们的社会、道德和宗教生活中的诸多因素，把我们分裂成了印度教徒、穆斯林、基督教徒、佛教徒等。各派宗教对于这场灾难负有很大的责任。我们也谈到了时间，时间就是过去——所有的记忆、积累的经验等——经过现在的调整然后延续到未来。这就是我们的生活。据生物学家和考古学家讲，我们在这个地球上已经存在了数百万年。在这个漫长的所谓进化过程中，我们积累了庞大的记忆。我们也探讨了记忆和思想的局限，以及这种局限是如何从地理上、国家上和宗教上分裂了这个世界的。

As we said the other day, this is not a lecture to inform or instruct you. You are not listening, if I may most respectfully point out, to a series of ideas or conclusions. But rather you and the speaker are taking a journey together like two friends talking over not only their family problems but also the world problems. So it is as much your responsibility to listen carefully, as it is that of the speaker to say things clearly, so that both of us understand what we are talking about. We are going to talk over together about order and disorder, pleasure, love, sorrow and death. These are all rather complicated problems which we all have to face in our daily lives, whether we are rich or poor. We have got to face this problem of existence.

正如我们前几天所说的那样，这不是一场向你灌输知识或者指导你的讲座。如果我可以怀着极大的敬意指出的话，你不是在听取一系列的观念或结论，而是你和讲话者像两个朋友一样一起踏上一段旅程，不仅一起探讨他们的家庭问题，还要探讨世界性的问题。所以我们双方肩负着同样重要的责任：你仔细地倾听，而讲话者清晰地表达，这样我们双方都能够理解我们所谈的内容。我们将一起来探讨秩序和混乱、快乐、爱、悲伤和死亡这些问题。这些都是相当复杂的问题，无论我们富有还是贫穷，都需要在日常生活中面对这些问题。我们必须面对这个生存的问题。

Our lives, our daily lives are in disorder. Which means contradiction—saying one thing and doing another, believing in something but actually moving in a direction totally different from what we believe. This contradiction creates disorder. I wonder whether we are at all aware of this problem. Apparently, there is going to be more and more disorder in the world on account of bad governments and economic and social conditions. There is always the threat of war, which is becoming more and more imminent, more and more pressing, and governments all over the world, even the tiniest nations, are buying armaments. So throughout the world there is great disorder. And our daily lives are also in disorder, though we are always pursuing order. We want order, because without order human beings will inevitably destroy themselves.

我们的生活，我们的日常生活处于混乱中。这意味着矛盾——说一套做一套，相信一套，但实际上却在另一个完全不同的方向去行动，而这种矛盾制造了混乱。我想知道我们究竟有没有发觉这个问题。显而易见，因为糟糕的政府和恶劣的经济社会条件，这个世界上的混乱将会越来越严重。战争的威胁始终存在，而且变得越来越迫在眉睫，越来越步步紧逼，而全世界的政府，即使是最小的国家，都在购买武器。所以说全世界都有着巨大的混乱。而我们的日常生活也处于混乱之中，尽管我们一直在追求秩序。我们希望有秩序，因为没有秩序，人类就必然会毁灭自己。

I hope we are sharing this question, thinking together, observing together, listening together, having a dialogue in which you are participating. It is not a matter of gathering a few ideas and conclusions, but together we must find out why we live in disorder and whether there can be total order in our lives and therefore in society.

我希望我们是在一起分担这个问题，一起思考，一起观察，一起聆听，进行一场你也在参与的对话。这不是一件收集几个新理念、新结论的事情，而是我们必须一起搞清楚我们为什么生活在混乱中，以及我们的生活中进而社会中能不能拥有彻底的秩序。

Society is brought about by us. It is put together by us, by our greed, our ambition, our envy, and the concept of individual freedom. This sense of individuality has brought about a great deal of disorder. Please, we are not attacking anything; we are merely observing what is going on in the world. In our lives, as we live it now after all these million years, there is still disorder. And we have always sought order, because without order we cannot possibly function freely, holistically. So we must find out what is order—not a blueprint, not something we put into a framework and follow. Order is something that is active, living; it does not conform to a pattern, whether the pattern be idealistic, historical or dialectical conclusions, or religious sanctions. Religions throughout the world have laid down certain laws, certain sanctions, and commandments, but human beings have not followed them at all. So we can put aside all those ideological conclusions and religious beliefs, which have nothing to do with our daily existence. We may conform to and follow certain laws laid down by religions, but that only brings about great bigotry, and so on.

社会是我们造就的，是由我们自己，由我们的贪婪、我们的野心、我们的嫉妒及个人自由的理念建立起来的。这种意义上的个体性带来了大量的混乱。请注意，我们并不是在攻击任何事情，我们只是在观察实际上正在发生的事情。在我们的生活中——过了几百万年，我们还是像现在这样生活着——依然存在着混乱。而我们一直在寻求秩序，因为没有秩序，我们就不可能自由地、完整地运转。所以我们必须发现什么是秩序——但不是蓝图，不是我们放入镜框架然后遵循的东西。秩序是一种鲜活的、有生命力的东西；它不会遵照某个模式，无论那个模式是理想的、历史的还是辩证的结论，抑或是宗教的条规。全世界的宗教都设下了某些戒律、某些条规和命令，但人类根本就没有遵守它们。所以我们可以把所有那些意识形态上的结论和宗教信仰都抛在一边，它们与我们的日常生活毫无关系。我们可以遵守和服从宗教定下的某些律条，但那只会带来严重的偏执，等等。

What is order? Is it possible to find out what is order when our brains are confused and disorderly? So we must first find out what is disorder, not what is order. Because when there is no disorder, there will naturally be order. Right? One of the causes of this disorder, perhaps the major cause, is conflict. Where there is conflict—not only between man and woman but also between nations, between religious beliefs and faiths—there must be disorder. Another major cause of disorder is this concept, this illusion that we are all individuals. As we said in the previous talks, you must question, doubt what the speaker is saying. Do not accept anything from anybody, but question, investigate, and not resist. If you merely resist what is being said, which may be true or false, then our conversation comes to an end. When two people are talking over together their problems, and if one is resisting, then conversation ends.

秩序是什么？当我们的大脑困惑、混乱时，有可能发现秩序是什么吗？所以我们必须首先搞清楚混乱是什么，而不是秩序是什么。因为当混乱不存在时，秩序自然就会出现，对吗？这种混乱的原因之一，也许是最主要的根源，就是冲突。当冲突存在时——不仅仅是男人和女人之间的冲突，还有国家之间、宗教信仰和信条之间的冲突——混乱就必然存在。混乱的另一个主要原因是这个观念、这个幻觉：我们都是个体。正如我们在之前的讲话中提到的那样，你必须质疑、怀疑讲话者所说的话。不要接受任何人说的任何事情，而是要质疑、探究，但不是抗拒。如果你单纯地抗拒我说的话，无论这些话是真是假，那么我们之间的对话就结束了。当两个人一起探讨他们的问题，如果其中一个人心怀抗拒，那么对话就结束了。

The major causes of disorder in our lives—each one thinking he is free, each one thinking of his own fulfilment, his own desires, his own ambitions, his own private pleasures. We are going to find out whether individuality is a fact or a long-established, respectable illusion. May we go into this together without accepting or denying? It is foolish to say,“I agree with you”or“I disagree with you”. You don’t “agree”or“disagree” with the sunrise and the sunset; it is a fact. You never say "I agree with you that the sun rises in the east". So could we perhaps put aside the sense of agreement and disagreement and inquire, without any bias or resistance, into whether there is actually individuality or whether there is something entirely different?

我们生活混乱的一个主要原因在于每个人都以为自己是自由的，每个人都想着他自己的成就、自己的欲望、自己的野心、自己的快乐。我们这就来搞清楚个体性是一个事实，还是一个根深蒂固的、体面的幻觉。我们可以一起探究这个问题，既不接受也不拒绝吗？说“我同意你的看法”或者“我不同意你的看法”，是很愚蠢的。你不会“同意”或者“不同意”日出和日落，那是一个事实。你从不会说“我同意你说的太阳会从东方升起”。所以，我们能不能抛开同意和不同意之类的感觉，不带有任何偏见和抗拒，一起来探询个体性是不是真的存在，还是说存在着某种截然不同的东西？

Our consciousness is the result of millions of years or more. It contains all the animalistic, primitive essence, as we have come from the animal, from nature. Deep down in our consciousness we find that there are still the deep responses of the animals, the fears, and the desire for security. All that is part of our consciousness. Our consciousness also contains innumerable beliefs, faiths, reactions, actions, various memories, fear, pleasure, sorrow, and the search for complete security. All that is what we are. We may think part of us is divine, but that is also part of our thinking. All that consciousness, we think, belongs to each one of us. Right? Religions—Christianity, Hinduism, and others—have maintained that we are separate souls.

我们的意识是数百万年或者更久时间的结果，它包含了所有动物性的、原始的要素，因为我们是从动物、从自然中进化而来的。我们发现，在我们的意识深处依然存在着那些深层的动物反应，存在着恐惧和对安全的渴望。这一切都是我们意识的一部分。我们的意识还包括不计其数的信仰、信念、反应、行动，包括各种记忆、恐惧、快乐、悲伤及对彻底安全的追寻。我们就是这一切。我们也许认为我们身上有一部分是神圣的，但那也是我们思想的一部分。我们以为那整个意识分属我们每一个人，对吗？各派宗教——基督教、印度教以及其他宗教——都声称我们是各自分离的灵魂。

Now, we are questioning the whole of it. Do you not share the sorrow of the rest of human beings? Human beings throughout the world have various forms of fear and various forms of pleasure. They suffer as you suffer. They pray, they do all kinds of absurd ceremonies as you do, seeking stimulation, sensation through ceremonies as you do. So you share the consciousness of all humanity; you are the entire humanity. First see it logically. Every human being on this earth, whatever be his religion or belief, suffers. Every human being suffers, deeply or superficially, and tries to evade suffering. So this consciousness, which we have considered"mine", this"personal consciousness"is not a fact. Because all human beings living on this marvellous, beautiful earth—which we are carefully destroying—go through the same problems, the same pain, anxiety, loneliness, depression, tears, laughter, contradiction, and conflict between man and woman, husband and wife. So in your consciousness are you individuals? Because that is what you are: your consciousness. Whatever you think or imagine, whatever your tendencies, aptitudes, talents or gifts, all that is shared by all other human beings, who are exactly as you are, similar to you. This is a logical fact. And logic has a certain place, one must think clearly, logically, reasonably, sanely. But logic is based on thought. However logical one may be, thought is limited; thought may be reasonable but it is limited. So one must go beyond thought, beyond the limitation of reason and logic.

现在我们就来质疑这一切。你难道没有和其他人类感受同样的悲伤吗？全世界的人类都有各式各样的恐惧和各式各样的快乐。他们和你一样遭受着痛苦。就像你一样，他们祈祷，他们执行各种各样荒唐的仪式，他们像你一样通过各种仪式寻求刺激和感官享受。所以你和全人类共有一个意识，你就是整个人类。首先理性地看到这一点。这个地球上的每一个人，无论他的宗教或者信仰是什么，都受着苦。每个人都或深或浅地遭受着痛苦，然后试图避开痛苦。所以，这个意识——我们认为它是属于“我的”——这个“个人的意识”

并非事实。因为生活在这个神奇而美丽的地球上的所有人类——我们正在精心地破坏着这个地球——经历着同样的问题，同样的痛苦、焦虑、孤独、沮丧、眼泪、欢笑、矛盾，以及男人与女人、丈夫与妻子之间的冲突。所以说在你的意识里，你能算是个体吗？因为那就是实际的你：你的意识就是你。无论你思考或者想象什么，无论你的倾向、能力、天赋或者天分是什么，这一切都为其他所有人类所共有，他们和你完全一样，和你极其相似。这是一个符合逻辑的事实。而逻辑具有一定的地位，你必须清晰地、符合逻辑地、理智地、清醒地思考。但逻辑是基于思想的。无论一个人多么理性，思想都是局限的；思想也许是理智的，但它是局限的。所以你必须超越思想，超越理性和逻辑的局限。

So you are the entire humanity; you are not an individual. Listen to that statement: you are the entire humanity, you are humanity, not all that rubbish of division. When you listen to a statement of that kind, do you make an abstraction of it? That is, do you make an idea of the fact? The fact is one thing, and the idea about the fact another. The fact is that you have thought that you are individuals. Your religions, your daily life, your conditioning have made you believe that you are individuals. And somebody like the speaker comes along and says,"Look carefully, is that so?"First you resist it saying, "What are you talking about?"and push it aside. But if you listen carefully, then you share this statement that you are the entire humanity. How do you hear that statement—the sound of it? Do you make out of that statement an idea, away from the fact, and pursue the idea?

所以你就是全体人类，你并不是一个个体。听听这个表述：你就是全体人类，你就是人类，而不是所有那些形同垃圾的划分。当你听到这样一种表述，你会把它抽象化吗？也就是说，你会把事实变成一种理念吗？事实是一回事，而对于事实的理念是另外一回事。事实是你一直认为自己是个体。你的宗教，你的日常生活，你所受的制约让你相信你是个体。然后像讲话者这样的一个人过来说："仔细瞧瞧，是这么回事吗？"你先是会抗拒，你会说："你在说什么啊？"然后把这个说法搁在一边。然而，如果你仔细倾听，那么你就是在分享"你就是全体人类"这个表述。你是如何听到这个表述——听到它的声音的？你会从这个表述中得出一个脱离事实的理念，然后再去追逐这个理念吗？

You hear the statement that your consciousness, with all its reactions and actions, is shared by all humanity, because every human being goes through desperation, loneliness, sorrow, and pain as you do. How do you listen to that statement? Do you reject it, or do you examine it? Do you investigate it or merely say"What nonsense"?

Which is it that you are doing—not tomorrow but now? What is your reaction to it? Either you listen to the depth of it, the sound of it, the beauty of it, the immensity of it, with its tremendous responsibility, or you treat it superficially, verbally, and say,"Yes, I understand it intellectually."Intellectual comprehension has very little meaning. It must be in your blood, in your guts, and out of that comes a different quality of the brain that is holistic, not fragmentary. It is the fragment that creates disorder. We, as individuals, have fragmented the human consciousness and therefore we live in disorder.

你听到了这个表述：你的意识，连同它所有的反应和行动，为全体人类所共有，因为每个人都经历着绝望、孤独、悲伤和痛苦，就像你一样。你是怎样去听这个表述的？你是拒绝它，还是检视它？你是探究它，还是只说一句“真是胡说八道”就完了？你做的是哪一样——不是明天，而就是现在？你对它的反应是什么？你要么聆听它的深度、它的声音、它的美、它的无限，以及它所包含的巨大责任，要么从表面上、从口头上去对待它，然后说，“是的，我从道理上明白了”。智力上的理解没什么意义。它必须进入你的血液、你的骨髓，从中就可以产生一个截然不同的大脑，它是完整的，毫无分裂的。正是分裂导致了混乱。我们作为个体，已经分裂了人类的意识，因此我们生活在混乱之中。

When you realize that you are the entire humanity—that is what love is. Then you will not kill another, you will not harm another. You will move out of all aggression, violence, and the brutality of religions. So your consciousness is one with all humanity. You don't see the beauty of it, the immensity of it. You will go back to your own pattern, thinking that you are all individuals, fighting, striving, competing, each trying to fulfil your own beastly little self. Yes, sir, this means nothing to you because you will go back to your own way of life. So it is much better not to listen to all this. If you listen to truth and don't act on it, it acts as poison. That is why our lives are so shoddy and superficial.

当你认识到你就是全体人类——那就是爱。那么你就不会去杀害别人，你也不会去伤害别人。你会摆脱所有的攻击性、暴力以及宗教的残酷。所以说，你的意识和所有人类是一体的。但是你没有看到它的美、它的无限。你还会回到你自己的模式中去，以为你们都是个体，不停地斗争、奋斗、竞争，每个人都试图满足自己那渺小无比的自我。是的，先生，这对你来说毫无意义，因为你还会回到你自己的生活方式中去。所以你大不如不来听这些讲话。如果你听到真理却不行动，那么它就是毒药。这就是为什么我们的生活会如此肤浅卑劣的原因。

We must also talk over together why man perpetually seeks pleasure—pleasure in possession, pleasure in achievement, pleasure in power, pleasure in having a status. There is sexual pleasure, which is maintained by constant thinking about sex, imagining, picturing, and making images. That is, thought gives pleasure; sensation is turned into pleasure. So we must understand what pleasure is and why we seek it. We are not saying it is right or wrong. We are not condemning pleasure, as we are not condemning desire. Desire is part of pleasure. The fulfilment of desire is the very nature of pleasure. Desire may be the cause of disorder—each one wanting to fulfil his own particular desire.

我们也必须一起来探讨为什么人永无休止地寻求快

感——占有之中的快感，成就之中的快感，权力之中的快感，拥有地位的快感。还有性快感，它由对性不停地思考、想象、描绘和制造意象这些活动所维系。也就是说，思想带来了快感，感官享受被转化成了快感。所以我们必须了解快感是什么，以及我们为什么要追求它。我们并没有说它是错是对。我们没有谴责快感，就像我们也没有谴责欲望一样。欲望是快感的一部分，欲望的满足正是快感的本质。欲望或许就是混乱的根源——每个人都想满足自己特定的欲望。

So together we are going to investigate whether desire is one of the major causes of disorder; we must explore desire, not condemn it, not escape from it, not try to suppress it. Most religions have said,“Suppress desire”—which is absurd. So let us look at it. What is desire? Put that question to yourself. Probably most of us have not thought about it at all. We have accepted it as a way of life, as the natural instinct of a man or a woman, and so we say, “Why bother about it?”Those people who have renounced the world, those who have entered monasteries, and so on try to sublimate their desires in the worship of a symbol or a person. Please bear in mind that we are not condemning desire. We are trying to find out what is desire, why man has, for millions of years, been caught not only physically but also psychologically in the trap of desire, in the network of desire.

所以，我们将一起来探究欲望是不是混乱的主要根源之一；我们必须探索欲望，而不是谴责它、逃避它，也不要试图压抑它。大部分宗教都说过“压制欲望”——那很荒唐。所以让我们来看看它。欲望是什么？问问自己这个问题。也许我们大多数人根本没想过这个问题。我们把它作为一种生活方式，作为男人或女人自然的本能接受了下来，所以我们说：“为什么要费心考虑它呢？”那些抛弃了世界、进入修道院之类地方的人，试图通过膜拜某个象征或者某个人，来升华他们的欲望。请谨记在心，我们并非在谴责欲望。我们在试着弄清楚欲望是什么，以及为什么数百万年以来，人类无论身心都被困在欲望的陷阱、欲望之网中。

Are you investigating with the speaker, or are you just listening to him while he explores or explains? You know, it is easy to be caught in explanations, in descriptions, and we are satisfied with the commentaries, descriptions, and explanations. We are not doing that here. I have to explain, describe, point out, put it in the framework of words, but you have to go into it too, and not just say,“Yes, I agree”, or “I disagree.” You have to find out the nature of desire, its construction, how it is put together, and what its origin is.

你是在和讲话者一起探究吗？还是说，他探索或者解释的时候你只是在听他说话而已？你知道，我们很容易就会被困在解释和描述中，我们满足于注释、描述和解说。我们在这里并没有这样做。我不得不解释、描述、指出，不得不诉诸语言，但你必须也同时进行探究，而不能只是说，“是的，我同意”或者“我不同意。”你必须弄清楚欲望的本质和结构，它是如何产生的，它的起源是什么。

The speaker will describe, not analyse. There is a difference between analysis and perception. Analysis implies the analyser and the thing to be analysed. Which means, the analyser is different from the analysed. But are they different? Suppose I am the analyser and I am envious. I begin to analyse why I am envious, as though I am different from envy. But envy is me; it is not separate from me. Greed, competition, comparison—all that is me. So we are not analysing; we are looking, hearing, and learning.

Learning is not merely accumulating memory. That is necessary, but learning is something entirely different. It is not an accumulative process. In learning, you are moving, fresh, never recording.

讲话者会描述，但不是分析。分析和洞察是不同的。分析意味着有分析者和被分析之物，也就是说，分析者与被分析之物是不同的。然而它们是不同的吗？假设我是分析者，而我很嫉妒。我开始分析我为什么嫉妒，就好像我和嫉妒是不同的。但嫉妒就是我，它和我是分不开的。贪婪、竞争、比较——这一切都是我。所以我们不是在分析；我们在看，在倾听，在学习。学习并非仅仅是累积记忆。记忆是必要的，但学习是完全不同的事情，它不是一个积累过程。在学习中，你一直在运动着，你始终是新鲜的，并且绝不会记录。

So we are observing desire, its origin, and why human

beings are caught in it endlessly. If you have a little money you want more. If you have a little power you want more. And power in any form, whether it be over your wife or your children, or political or religious power is an abominable thing. It is evil because it has nothing to do with truth. So what is the origin of desire? We live by sensation. If there was no sensation, biologically and psychologically, we would be dead human beings. Right? The cawing of that crow is acting on the eardrum and nerves, and the noise is translated as the cry of a crow. That is a sensation. Sensation is brought about by hearing or seeing, and then contact. You see a garden beautifully kept; the green is rich, perfect, and there are no weeds in it. It is a lovely thing to watch. There is the seeing. Then, if you are sensitive, you go and touch the grass. That is, seeing, contact, and then sensation. We live by sensation, it is necessary. If you are not sensitive, you are dull, you are half—alive, as most of us are. Take a very simple example. You see a nice sari or a shirt in a shop. You see it. You go inside and touch it; then there is the sensation of touching it and you say, “By Jove, what lovely material that is”. Then what takes place after that? Are you waiting for me to tell you? Please do listen to this. If you see this for yourself, and not be told by another, then you become the teacher and the disciple. But if you repeat, repeat, repeat what somebody, including the speaker, has said, you remain mediocre, thoughtless, repetitive. So let’s go into it.

所以我们是在观察欲望和它的起源，以及人类为什么会无休止地困在其中。如果你有一点儿钱，你就想得到更多的钱。如果你有一点儿权力，你就想拥有更多的权力。而任何形式的权力，无论是凌驾于你的妻子或孩子之上的权力，还是政治或宗教权力，都是一种极其可恶的东西。它是邪恶的，因为它与真理没有丝毫关系。那么欲望的起源是什么？我们依靠感官感受来生活。如果没有生理上和心理上的感受，我们就是死人，对吗？那只乌鸦呱呱的叫声作用在耳鼓和神经上，然后这个声音就被翻译成了乌鸦的叫声。这就是感官感受。感官感受由听觉、视觉和触觉产生。你看到一座精心打理的花园，满眼是完美的绿色，没有丛生的杂草，看上去真是一件赏心悦目的事。这是视觉。然后如果你敏感的话，你去触摸小草。也就是说，有了视觉、触觉，然后有了感受。我们依靠感官感受来生活，它是必不可少的。而如果你不敏感，你就会很迟钝，一副半死不活的样子，就像我们大多数人一样。举一个非常简单的例子：你看到商店里有一件漂亮的纱丽或者一件衬衫，你看到了它。你走进去摸了摸它，然后就有了触摸的感受，于是你说，“天哪，这料子可真不错”。接下来发生了什么？你在等我来告诉你吗？请务必好好听一听。如果你自己看到了这一点，而不是别人告诉你的，那么你就同时成了老师和弟子。

但是如果你重复、重复、重复别人说过的话，包括讲话者说过的话，你就会依然平庸、麻木、机械。所以让我们来深入地探讨一下这个问题。

You see a beautiful car, you touch the polish, see its shape and texture. Out of that there is sensation. Then thought comes and says,"How nice it would be if I got it, how nice it would be if I got into it and drove off."So what has happened? Thought has intervened, has given shape to sensation. Thought has given to sensation the image of you sitting in the car and driving off. At that moment, at that second, when thought creates the image of you sitting in the car, desire is born. Desire is born when thought gives a shape, an image, to sensation. Now, sensation is the way of existence, it is part of existence. But you have learnt to suppress, conquer, or live with desire with all its problems. Now, if you understand this, not intellectually but actually, that when thought gives shape to sensation, at that second desire is born, then the question arises: Is it possible to see and touch the car—which is sensation—but not let thought create the image? So keep a gap. Do you understand this?

你看到一部漂亮的汽车，你抚摸亮漆，看到了它的外形和质地，从中就产生了感官感受。然后思想过来说："如果我拥有了它，那该多好啊，如果我坐进去然后疾驰而去，那该多棒啊。"那么这里发生了什么？思想干涉了感受，赋予了它形状。思想把你坐在车里疾驰而去的意象，带给了感受。在那一刻，在那一秒钟，当思想制造出你坐在车里的意象时，欲望就产生了。当思想带给感受某个形状或者意象时，欲望就产生了。而感受是我们存在的方式，是生活的一部分。但是，你已经学会了压抑欲望、征服欲望，或者与它所有的问题生活在一起。那么，如果你明白了这一点，不是从智力上，而是真正地懂得了在思想赋予感受形状的那一瞬间，欲望就产生了，那么这个问题就出现了：有没有可能看到并抚摸那辆车——这是感受——却不让思想制造出意象？所以在这里保留间隙。这一点你明白了吗？

You see, one must also find out what is discipline. Because it is related to desire. The word“discipline”comes from the word“disciple”; the etymological meaning of that word is“one who is learning”. A disciple is one who is learning—learning, and not conforming, not controlling, not suppressing, not obeying, not following. On the contrary, he is learning from observation. So you are learning about desire. Learning about it is not memorizing. Most of you are trained, especially if you are in the army, to discipline yourselves according to a pattern, trained to copy, follow, obey. That is what you are all doing, hoping that discipline will bring about order. But if you are learning, then that very learning becomes its own order; you don't need order imposed by law or anything else. So learn, find out, whether it is possible to allow sensation to flower and to not let thought interfere with it—to keep them apart. Will you do it? Then you will find that desire has its right place. And when you understand the nature of desire, there is no conflict about it.

你看，我们也必须弄清楚戒律是什么，因为它和欲望是联系在一起的。“戒律”这个词来源于“弟子”这个词，而这个词在语源上的意思是“正在学习的人”。弟子是一个正在学习的人——学习，而不是遵守，不是控制，不是压抑，不是服从，不是追随。正相反，他从观察中学习。所以你正在学习有关欲望的一切，而这种学习不是记忆。你们大部分人受到的训练都是依照某个模式来约束自己，尤其是如果你在军队里的话，你被训练去模仿、跟随和服从。这就是你们都在做的事情，希望那种戒律能够带来秩序。但是如果你在学习，这种学习本身就会成为它自己的秩序，你不需要法律或者其他任何东西强加的秩序。所以去了解、去弄清楚有没有可能让感受绽放，而不让思想去干涉它——把它们分开。你愿意这么做吗？然后你就会发现欲望有了它自己正确的位置。而当你了解了欲望的本质，就不会再有关于欲望的冲突了。

We also ought to talk over together love, sorrow, and death. Please, all this is much too serious as it affects your daily life. This is not something you play with intellectually; it concerns your life—not somebody elses' life—the way you live after all these million years. Look what your lives are, how empty, shallow, violent, brutal, inconsiderate, thoughtless. Look at it. All this has created such havoc in the world. You all want to have high positions, achieve something, become something. And seeing all this, there is great sorrow, isn't there? Every human being in this world, whether he is highly placed or is just an ignorant villager, goes through great sorrow. He may not recognize the nature and the beauty and the strength of sorrow, but he goes through pain like you do. Human beings have gone through sorrow for millions of years. They haven't solved the problem; they want to escape from it. And what is the relationship of sorrow to love and death? Can there be an end to sorrow? This has been one of the questions mankind has asked for millions of years. Is there an end to all the pain, the anxiety, the grief of sorrow?

我们也应该一起来探讨爱、悲伤和死亡。请注意，这一切都是太严肃的话题了，因为它们影响着你的日常生活。这不是一场智力游戏，这与你的生活紧密相关——不是别人的生活——而是你几百万年以来的生活方式。看看你的生活是什么，它是多么空虚、浅薄、暴力、残酷、冷漠、自私。看看它。这一切在世界上造成了如此严重的破坏。你们都希望坐拥高位，功成名就。而看到了这一切，你就会感到巨大的悲伤，不是吗？这世上的每一个人，无论身居高位或是一个愚昧的村民，都经历着巨大的悲伤。他也许意识不到悲伤的本质、美和力量，但是他像你一样经历着痛苦。人类已然历经了数百万年的悲伤，他们没有解决这个问题，他们想要逃避它。而悲伤与爱和死亡又有着怎样的关系？悲伤能够终结吗？这是人类问了数百万年的问题之一。这所有的痛苦、焦虑和沉痛的悲伤是否有尽头？

Sorrow is not only your own particular sorrow; there is also the sorrow of mankind. Historically, there have been five thousand years of war. That means every year some people kill others for the sake of their tribe, their religion, their nation, their community, their individual protection, and so on. Have you ever realized what wars have done, the havoc they have created? How many millions have cried, how many millions have been wounded, left without arms, without legs, without eyes, even without a face? You people don't know anything about all this. So is there an end to sorrow and all the pain therein? And what is sorrow? Don't you know sorrow? Don't you? Are you ashamed to acknowledge it? When your son or daughter or somebody whom you think you love is taken away, don't you shed tears? Don't you feel terribly lonely because you have lost a companion forever? We are not talking about death but about this immense thing that man goes through without ever having a solution.

悲伤不仅仅是你自己特有的悲伤，还有人类的悲伤。从历史上看，战争已经存在了五千年。这意味着每年都有一些人在杀害另一些人，为了他们的部落、他们的宗教、他们的国家、他们的团体、他们个人的安全，等等。你可曾意识到战争都做了些什么，它们制造了怎样的浩劫？你可知有几百万人在哭泣，有几百万人受了伤，失去了手臂、腿脚和眼睛，甚至连面孔也失去了？你们这些人根本不知道。那么悲伤及其中所有的痛苦能够终结吗？而悲伤又是什么？你们难道不了解悲伤吗？不了解吗？你们羞于承认这一点吗？当你的儿子、女儿或者你认为自己爱着的某个人被夺走了，你难道不会流泪吗？你难道不会因为永远失去了某个人的陪伴而感到极其孤独吗？我们现在谈的不是死亡，而是人类经历的这个从来都未能解决的庞然大物。

Without ending sorrow there is no love. Sorrow is part of your self-interest, part of your egotistic, self-centred activity. You cry for another, for your son, for your brother, for your mother. Why? Because you have lost something that you are attached to, something which gave you companionship, comfort, and all the rest of it. With the ending of that person, you realize how utterly empty, how lonely your life is. Then you cry. And there are many, many people ready to comfort you, and you slip very easily into that network, that trap, of comfort. There is the comfort in God, which is an image put together by thought, or comfort in some illusory concept or idea. And that's all you want. But you never question the very urge, the desire for comfort, never ask whether there is any comfort at all. One needs to have a comfortable bed or chair—that's all right. But you never ask whether there is any comfort at all psychologically, inwardly. Is it an illusion which has become your truth? You understand? An illusion can become your truth—the illusion that you are God, that there is God. That God has been created by thought, by fear. If you had no fear, there would be no God. God has been invented by man out of his fear, loneliness, despair, and the search for everlasting comfort. So you never ask whether there is comfort at all, which is deep, abiding satisfaction. You all want to be satisfied not only with the food that you eat, but also satisfied sexually, or by achieving some position of authority and having comfort in that position. So let us ask whether there is any comfort at all, whether there is anything that will be gratifying from the moment we are born till we die. Don't just listen to me; find out, give your energy, your thought, your blood, your heart to find out. And if there is no illusion, is there any comfort? If there is no fear, do you want comfort? Comfort is another form of pleasure.

悲伤若不终结，爱就无法存在。悲伤是你自私自利的一部分，是你自我本位和自我中心行为的一部分。你为另一个人哭泣，为你的儿子、你的兄弟、你的母亲哭泣，为什么？因为你失去了你依恋的东西，失去了带给你陪伴、舒适等诸如此类感觉的东西。那个人的死去，让你意识到你的生命是如此空虚、如此孤独，于是你会哭泣。而同时又有许许多多人愿意去安慰你，然后你就轻易地落入了那个舒适的罗网和陷阱中。神明之中有舒适感，而那只是思想制造出来的意象，或者某些虚幻的概念或想法中也有舒适感。而这就是你想要的一切。但是你从不质疑对舒适的这种渴望、这种需求本身，从不询问舒适究竟是否存在。你需要一张舒适的床或者椅子——这完全没问题。但你从不问心理上、内在的舒适究竟是否存在。它是不是成了你的真理的一个幻觉？你明白吗？一个幻觉可能会成为你的真理——“你就是神”“存在着神”这些幻觉。那个神是思想、是恐惧制造出来的。如果你没有恐惧，就根本不会有什么神明。神是人类出于恐惧、孤独、绝望及对永恒舒适的追求而虚构出来的。所以你从来不问舒适——那种深深的、持久的满足感——究竟是否存在。你们都想得到满足，不仅对你吃的食物满足，而且也要得到性满足，或者通过取得某种权威的地位并在那个位子上获得舒适和满足。所以让我们来问一问舒适究竟是否存在，让我们从生到死都感到满意的东西是否存在。不要只是听我讲；去搞清楚，付出你的能量、你的思想、你的热血、你的心去弄清楚。而如果没有幻觉，还会有任何舒适存在吗？如果没有恐惧，你还会想要舒适吗？舒适是另一种形式的快感。

So this is a very complex problem of our life—why we are so shallow, empty, filled with other people's knowledge and with books; why we are not as independent, free human beings to find out; why we are slaves. This is not a rhetorical question; it is a question each one of us must ask. In the very asking and doubting, there comes freedom. And without freedom there is no sense of truth.

所以说，这是我们的生活中一个非常复杂的问题——我们为什么如此浅薄、如此空虚，塞满了别人的知识和书本的知识；我们为什么不能独立、自由地去探究；我们为什么是奴隶。这并不是夸大其词的问题，而是我们每个人都必须提出的问题。在这种提问和质疑当中，自由就会到来，而没有自由，就没有对真理的感知。

So we will go tomorrow into the question of what is a religious life and whether there is something that is totally sacred, holy, something not invented by thought.

我们明天再来探讨什么是宗教生活这个问题，以及是否存在某种完全神圣、圣洁的东西，某种并非由思想发明的东西。

一九八五年二月九日

至福就在身边　TALK Ⅳ　讲话四

WE were talking yesterday evening about sorrow and the ending of sorrow. With the ending of sorrow there is passion. Very few of us really understand or go deeply into the question of sorrow. Is it possible to end all sorrow? This has been a question that has been asked by all human beings, perhaps not very consciously; but deeply they have wanted to find out, as we all do, if there is an end to human suffering, human pain and sorrow. Because without the ending of sorrow, there is no love. When there is sorrow it is a great shock to the nervous system, like a blow to the whole physiological as well as psychological being. We generally try to escape from it by taking drugs or drink or through every form of religion. Or we become cynical or accept things as being inevitable.

我们昨晚谈到了悲伤和悲伤的终结。伴随着悲伤的终结而来的是激情。我们很少有人真正理解了或者深入地探究过悲伤的问题。有可能终结所有的悲伤吗？这是所有人类都曾经问过的一个问题，只是也许并不是那么有意识地提出的；但是在内心深处，他们都想弄清楚人类的不幸、人类的痛苦和悲伤能否终结，就像我们所有人希望的一样。因为悲伤若不终结，就不会有爱。当悲伤存在时，它是对神经系统的一次巨大震撼，就像是对整个身心的一次重大打击。我们通常试图借助服用药物、饮酒或者各种形式的宗教来逃避悲伤。要么我们就变得愤世嫉俗，或者把一切都当作不可避免的接受下来。

Can we go into this question very deeply, seriously? Is it possible not to escape from sorrow at all? Perhaps my son dies, and there is immense sorrow, shock, and I discover that I am really a very lonely human being. I cannot face it, I cannot tolerate it. So I escape from it. And there are many escapes—mundane, religious, or philosophical. This escape is a waste of energy. Not to escape in any form from the ache, the pain of loneliness, the grief, the shock, but to remain completely with the event, with this thing called suffering—is that possible? Can we hold any problem—hold it and not try to solve it—try to look at it as we would hold a precious, exquisite jewel? The very beauty of the jewel is so attractive, so pleasurable that we keep looking at it. In the same way if we could hold our sorrow completely, without a movement of thought or escape, then that very action of not moving away from the fact brings about a total release from that which has caused pain. We will go into this a little later.

我们能不能非常深入而认真地探究一下这个问题？有没有可能根本不逃避悲伤？也许我的儿子去世了，我感受到了巨大的悲伤和打击，我发现我真的是一个非常孤独的人了。我无法面对、无法容忍这一点，所以我逃避它。而确实也有很多逃避的办法——世俗的、宗教的，或者哲学的。这种逃避是一种能量的浪费。不采取任何方式逃避孤独、悲伤和打击带来的痛苦，而是完全与那件事情，与那个叫作苦难的东西共处——这可能吗？我们能不能捧着问题——捧着它而不试图解决它——尝试去看着它，就像手捧一件精致的稀世珍宝？这珍宝本身的美是如此迷人、如此悦目，以至于我们一直看着它。同样，如果我们能够完全拥抱我们的悲伤，没有一点思想或者逃避的活动，那么这个不逃避事实的行动本身，就会带来一种彻底的解放，彻底摆脱了痛苦的肇因。我们稍后再来探讨这一点。

And we should also consider what is beauty. Beauty is very important—not the beauty of a person or of the marvellous paintings and statues in museums and ancient man's endeavour to express his feelings in stone or in paint or in poem. We should ask ourselves what is beauty. Beauty may be truth, beauty may be love. Without understanding the nature and the depth of that extraordinary word"beauty", we may never be able to come upon that which is sacred. So we must go into the question of what is beauty.

我们也应该来思考一下美是什么。美非常重要——不是一个人的美，不是博物馆里杰出的画作和雕塑的美，也不是古人为了在石头上、图画中或者诗歌中表达自己的感受所做的努力。我们应该问问自己美是什么？美也许就是真理，美也许就是爱。如果不理解“美”这个非凡的词的本质和深度，我们也许永远都无法遇到神圣。所以我们必须来探讨一下美是什么这个问题。

When you see something greatly beautiful like a mountain full of snow against the blue sky, what actually takes place? When you see something extraordinarily alive, beautiful, majestic, for a moment, for a second, the very majesty of that mountain, tne immensity of it drives away, puts aside all self-concern, all problems. At that second there is no“me”watching it. The very greatness of the mountain drives away for a second all my self-concern. Surely one must have noticed this. Have you noticed a child with a toy? He has been naughty all day long—which is right—and you give him a toy, and then for the next hour, until he breaks it, he is extraordinarily quiet. The toy has absorbed his naughtiness, has taken him over.

当你看到一样极其美丽的事物，比如蓝天掩映下白雪覆盖的高山，实际上会发生什么？当你看到了一件无比鲜活、美丽而又庄严的事物，在那一刻、那一秒钟，那高山的伟岸和无限驱散了，消除了所有的自私、所有的问题。那一秒钟并没有一个“我”在看着它。高山的伟岸本身就短暂地驱散了我所有的自私自利。你肯定曾经留意到了这一点。你有没有注意过一个有了玩具的孩子？他一整天都很淘气——这也没什么不好——而当你给了他一个玩具，然后在接下来的一个小时里他都非常安静，直到他把玩具弄坏。玩具消除了他的顽皮，完全占据了他。

Similarly, when we see something extraordinarily beautiful, that very beauty absorbs us. That is, there is beauty when there is no self, no self-interest, no travail of the self. Without being absorbed or shaken by something extraordinarily beautiful like a mountain or a valley in deep shadow, without being taken over by the mountain, is it possible to understand beauty that is without the self? Because where there is the self there is no beauty, where there is self-interest there is no love. So love and beauty go together; they are not separate.

同样，当我们看到极其美丽的事物，那美本身就完全吸引了我们。也就是说，当自我、自私自利、自我的痛楚不存在时，美就出现了。如果没有被极其美丽的事物，比如一座山或者幽深的峡谷所吸引、所震撼，没有被高山所占据，我们有可能理解没有自我的美吗？因为有自我就没有美，有自私自利就没有爱。所以爱和美是并肩而行的，它们不是分开的。

We have also to talk over together what is death. That is one certain thing that we all have to face. Whether we are rich or poor, ignorant or full of erudition, death is certain for every human being; we are all going to die. And we have never been able to understand the nature of death. We are always frightened of dying, aren't we? And we hope for continuity after death. So we are going to find out for ourselves what is death, because we are going to face it whether we are young or old. And to understand death, we must also inquire into what is living, what is our life.

我们也必须一起来谈谈死亡是什么。这是一件我们都必须面对的事情。无论我们富有还是贫穷，愚昧无知还是学富五车，死亡对每个人来说都是必然发生的事情，我们都会死。而我们从来都没能了解死亡的本质。我们总是害怕死亡，不是吗？我们希望死后生命能够延续。所以我们要自己去弄清楚死亡是什么，因为无论我们年轻还是年老，我们都要面对它。而若要了解死亡，我们也必须探询什么是活着，我们的生命是什么。

Are we wasting our lives? By that word“wasting”we mean dissipating our energy in various ways, dissipating it in specialized professions. Are we wasting our whole existence, our life? If you are rich you may say, “Yes, I have accumulated a lot of money, it has been a great pleasure.”Or if you have a certain talent, that talent is a danger to a religious life. Talent is a gift, a faculty, an aptitude in a particular direction, which is specialization. Specialization is a fragmentary process. So you must ask yourself whether you are wasting your life. You may be rich, you may have all kinds of faculties, you may be a specialist, a great scientist or a businessman, but at the end of your life has all that been a waste? All the travail, all the sorrow, all the tremendous anxiety, insecurity, the foolish illusions that man has collected, all his gods, all his saints and so on—have all that been a waste? You may have power, position, but at the end of it—what? Please, this is a serious question that you must ask yourself. Another cannot answer this question for you.

我们是在浪费自己的生命吗？用“浪费”这个词，我们的意思是指以各种方式，在各种专门化的职业中耗散我们的能量。我们是在浪费我们的整个生命、我们的生活吗？如果你很富有，你也许会说：“是的，我积攒了很多钱财，这是一种巨大的快感。”或者，如果你有某种天赋，那天赋对宗教生活来说就是一种危险。天赋是在某个特定的方向上具有的天分、才能和天资，那是一种专门化，而专门化是一个支离破碎的过程。所以你必须问问你自己是不是在浪费自己的生命。你也许很富有，也许拥有各种才能，也许是个专家，是个大科学家或者是个商人，但是，在你生命的尽头，那一切是不是一种浪费？所有的痛楚、所有的悲伤、所有巨大的焦虑感、不安全感，人类搜集的各种愚蠢的幻觉，他所有的神明、所有的圣人等——这一切是不是一种浪费？你也许拥有权力、地位，但是到了最后——又能怎样呢？拜托，这是一个很严肃的问题，你必须问问你自己，别人无法替你回答这个问题。

So we have separated living from dying. The dying is the end of our life. We put it as far away as possible—a long interval of time—but at the end of the long journey we die. And what is it that we call living? Earning money, going to the office from nine to five, over-worked either in a laboratory or in an office or in a factory, and the endless conflict, fear, anxiety, loneliness, despair, depression—this whole way of existence is what we call life, living. And to that we hold. But is that living? This living is pain, sorrow, anxiety, conflict, every form of deception, and corruption.

Where there is self-interest there must be corruption. This is what we call living. We know that, we are very familiar with all that, that is our daily existence. And we are afraid of dying, which is to let go of all the things that we have known, all the things that we have experienced and gathered—the lovely furniture and the beautiful collection of pictures and paintings. And death comes and says,"You cannot have any of those things any more."So we cling to the known, afraid of the unknown.

所以我们把生与死分离了开来。死亡就是我们生命的结束，我们竭尽所能把它推得远远的——推到一段漫长的时间之后——然而，就在这段漫长旅程的尽头，我们还是会死去。而什么又是那个我们叫作生活的东西呢？赚钱，朝九晚五地上班去，要么在实验室，要么在办公室或者工厂里加班加点，还有无尽的冲突、恐惧、焦虑、孤独、绝望和沮丧——这整个生存方式，我们称之为生活、生存，并且紧紧抓住不放。然而这是生活吗？这种生活就是痛苦、悲伤、焦虑、冲突，是各种形式的欺骗和腐败，而哪里有自私自利，哪里就必然会有腐败。这就是我们叫作生活的东西。我们知道这些，我们对这一切都非常熟悉，这就是我们的日常生活。同时我们也害怕死亡，死亡就是放开我们已知的一切，我们经历过和积攒起来的一切——漂亮的家具，收藏的美丽照片和画作。然后死亡来临并且说："所有这些东西你都不能再拥有了。"所以我们执着于已知，害怕未知。

We can invent reincarnation. But we never inquire into what it is that is born next life. What is born next life is a bundle of memories. Because we live by memories. We live by the knowledge we have acquired or inherited, and that knowledge is what we are. The self is the knowledge of the past experiences, thoughts, and so on. The self is that. The self may invent something divine in one. But it is still the activity of thought, and thought is always limited. So this is our living, this is what we call life—pleasure and pain, reward and punishment. And death means the ending of all that, the ending of all the things that we have thought, accumulated, enjoyed. And we are attached to all that. We are attached to our families, to all the accumulated memory, to knowledge, to the beliefs and the ideals we have lived with. We are attached to all that. And death says ,"That's the end of it, old boy."

我们可以编造出轮回转世的说法，但我们从不探究来世再生的是什么。来世再生的是一团记忆，因为我们依靠记忆为生。我们依靠后天获取的或者先天继承来的知识为生，而那知识就是我们自己。自我就是过去的经验、思想等的知识，自我就是这些。自我也许会在人类身上虚构出某种神圣的东西，但那依然是思想的活动，而思想始终是局限的。所以这就是我们的生活，这就是我们所谓的生活——快乐和痛苦，奖励和惩罚。而死亡意味着这一切的终结，我们所思所想、所积累、所享受的一切的终结。我们依附于这一切，我们依附于我们的家庭，依附于积累起来的所有记忆、知识、信仰和理想，我们就与它们生活在一起。我们依附于这一切。而死亡说：“到此结束了，老兄。”

Now, the question is: Why has the brain separated living—living which is conflict and so on—from death? Why has this division taken place? Does this division exist when there is attachment? Please, we are talking over things together, we are sharing this thing which man has lived with for millions of years—the living and the dying. So we have to examine it together, and not resist, not say, “Yes, I believe in reincarnation, that’s what I live by, to me that is important.”Otherwise the conversation between us will come to an end. So we should really go into the question of what is living, what is wasting one's life, and what is dying. One is attached to so many things—to the guru, to accumulated knowledge, to the memory of one's son or daughter, and so on. That memory is you. Your whole brain is filled with memory—memory not only of recent events but also the deep abiding memory of that which has been the animal, the ape. We are part of that memory. We are attached to this whole consciousness. That's a fact. And death comes and says,"That is the end of your attachment."So we are frightened of that, frightened of being completely free from all that. And death is that—the cutting off of everything that we have got.

接下来我们的问题就是：大脑为什么把生与死分开？生活中为什么会有这种划分？是不是有依附的时候这种划分才会存在？请注意，我们是在一起探讨问题，我们是在分享这件陪伴了人类数百万年的事情——生与死。所以我们必须一起来审视，不要抗拒，不要说，“是的，我相信轮回转世，这是我赖以为生的信念，这对我来说很重要”。否则我们之间的对话就结束了。所以我们真的应该一起来探讨这些问题：生活是什么，什么是浪费生命，以及死亡是什么。你依附于如此之多的事物——依附于古鲁，依附于积累起来的知识，依附于你对自己儿女的记忆，等等。那记忆就是你。你的整个大脑都装满了记忆——不仅仅是最近发生的事情的记忆，而且还有从动物、从猿类开始就有的深藏的持久的记忆。我们就是记忆的一部分，我们依附于这整个意识。这是一个事实。而死亡过来说，“你的依附到此结束”。所以我们害怕结束，害怕彻底脱离这一切。而死亡就是这个——切断我们已有的一切。

We can invent and say, “I will continue next life.” But what is it that continues? You understand my question? What is it in us that desires to continue? Is there a continuity at all, except of your bank account, your going to the office every day, your routine of worship, and the continuity of your beliefs? They are all put together by thought. And the self, the“me”, the ego, the persona, is a bundle of complicated ancient and modern memories. You can see it for yourself. You don’t have to study books and philosophies about all that. You can see for yourself very clearly that you are a bundle of memories. And death puts an end to all that memory. Therefore one is frightened. Now, the question is: Can one live in the modern world with death? Not suicide—we are not talking about that. But can you, as you live, end all attachment, which is death? I am attached to the house I am living in, I have bought it, I have paid a great deal of money for it, and I am attached to all the furniture, the pictures, the family, the memories. And death comes and wipes all that out.

我们可以编造说，“我来世还会再继续”。但继续的又是什么呢？你明白我的问题吗？我们身上的什么东西渴望延续？除了你的银行账户，你每天去办公室上班，你例行公事的敬拜和你信仰的延续之外，究竟存在什么延续性吗？这些都是思想拼凑出来的东西。而自我、“我”、自己、人格，是一捆古老和现代的复杂记忆。这一点你自己就能看清楚。你不需要去研读有关的书籍和哲学，你自己就可以非常清楚地看到你就是一捆记忆。而死亡会终结这所有的记忆，所以你感到害怕。那么，我们的问题就是：一个人能不能伴着死亡活在当今的世界上？不是自杀——我们说的不是这个，而是你能不能在活着的时候就终结所有的依附，也就是死去？我依附于我所居住的房子，我买了它，我为它花了好多钱，我还依附于所有的家具、照片、家庭和记忆。而死亡来了就把这一切都彻底摧毁了。

So can I live every day of my life with death? Ending

everything every day, ending all your attachments—that's what it means to die. But we have separated living from dying. Therefore we are perpetually frightened. But when you bring life and death together, the living and the dying, then you will find out that there is a state of the brain in which all knowledge as memory ends. But you need knowledge to write a letter, to come here, to speak English, to keep accounts, to go home, and so on. You need knowledge but not knowledge as something that entirely occupies the mind. We were talking the other day with a computer expert. The computer can be programmed and it stores that memory. It can also put aside all that memory on paper or a disc and keep itself empty so that it can be reprogrammed or instructed further. Similarly, can the brain use knowledge when necessary but be free of all knowledge? That is, our brain is recording all the time. You are recording what is being said now, and this record becomes a memory. That memory, that recording, is necessary in a certain area. That area is physical activity. Now, can the brain be free so that it can function totally in a different dimension? That is, every day, when you go to bed, wipe out everything that you have collected, die at the end of the day.

所以，我能不能每天都和死亡生活在一起？每天都结束一切，结束你所有的依附——死去就是这个意思。但我们把生与死划分开来，所以我们永远都觉得恐惧。然而，当你把生与死，把生命和死亡合而为一，你就会发现大脑出现了一种状态，其中所有的知识，也就是记忆，都结束了。但是你需要知识来写信，到这里来、说英语、记账、回家等都需要知识。你需要知识，但不是完全占满心灵的知识。我们前几天和一个计算机专家探讨过，计算机可以被程式化，它可以储存那些记忆。它也可以把所有那些记忆储存在纸上或者磁盘上，这样就能清空它自己，然后再重新被程式化或者接受进一步的指令。同样，大脑能不能在必要的时候使用知识，同时又能摆脱掉所有知识的束缚？也就是说，我们的大脑一直在记录，你在记录我现在所说的话，而这记录会变成记忆。这记忆、这记录在某个领域是必要的，这个领域就是身体上的活动。然而，大脑能不能获得自由，这样它就能在另一个维度上完整地运转了？也就是说，每一天，在你上床睡觉的时候，抹掉你收集起来的一切，在一天结束的时候死去。

You hear a statement of this kind—that is, living is dying; they are not two separate things at all. You hear the statement not only with the hearing of the ear, but if you listen carefully, you also hear the truth of it, the actuality of it, and for the moment you see the clarity of it. Later on, you slip back—you are attached, you know all the rest of it. So is it possible for each one of you to die at the end of the day to everything that is not necessary, to every memory of hurt, to your beliefs, your faiths, your anxieties, your sorrow? End all that every day and then you will find that you are living with death all the time—death being the ending. One should also go into the question of ending. We never end anything completely. We end if there is some profit in it, if there is some reward. Can we voluntarily end without the assumption that there is something better in the future? And it is possible to live that way in the modern world. That is a holistic way of living in which there is the living and the dying taking place all the time.

你听到了这样的一个表述——那就是，生就是死，它们完全不是两件分开的事情。你不仅仅用耳朵听到了这句话，而且如果你认真倾听的话，你也听到了它包含的真理和真实性，此刻你清楚地看到这一点。可是过后你又退回去了——你照样依附，你知道诸如此类的一切。所以，你们每一个人有没有可能在每天结束的时候，对不必要的一切死去，对每个伤害的记忆死去，对你的信仰、你的信念、你的焦虑、你的悲伤死去？每天都终结这一切，然后你就会发现你一直与死亡一起生活——死亡就是终结。我们也应该来探究终结这个问题。我们从来没有彻底地终结过任何事情。如果有利可图，如果有某种奖赏，我们才会结束一件事情。在不假定未来有某种更好的东西的情况下，我们能不能自愿地终结一件事情？我们有可能这样活在现代世界中吗？这是一种完整的生活方式，其中生与死一直是在同时发生的。

Then we ought to also talk over together what is love. Is love sensation? Is love desire? Is love pleasure? Is love put together by thought? Do you love your wife or your husband or your children—love? Is love jealousy? Don't say"no"because you are jealous. Is love fear, anxiety, pain, and all the rest of it? So what is love? You may be very rich, you may have power, position, importance, all that hierarchical outlook on life, but without love, without that quality, that perfume, that flame, you are just an empty shell. If you loved your children, would there be wars? If you loved your children, would you allow them to maim themselves through wars, kill others, hurt another? Can love exist where there is ambition? Please, you have to face all this. But you don't because you are caught in a routine, in a repetition of sensation as sex, and so on. Love has nothing whatsoever to do with pleasure, with sensation. Love is not put together by thought. Therefore it is not within the structure of the brain; it is something entirely outside the brain. While the brain by its very nature and structure is an instrument of sensation, nervous responses, and so on, love cannot exist where there is mere sensation. Memory is not love.

接下来我们也应该一起来探讨一下爱是什么。爱是感官享受吗？爱是欲望吗？爱是欢愉吗？爱是思想拼凑出来的吗？你爱你的妻子、你的丈夫或者你的孩子们吗——你爱吗？爱是嫉妒吗？不要说“不是”，因为你就在嫉妒。爱是恐惧、焦虑、痛苦以及诸如此类的一切吗？那么什么是爱呢？以所有那些等级化的生活观来看，你也许很富有，你也许有权有势，你也许很重要，然而如果没有爱，没有那种品质、那种芬芳、那种火焰，你就只是一具空壳。如果你爱你的孩子，还会有战争吗？如果你爱你的孩子，你会允许他们上战场把自己弄成残废，去杀害别人，去互相伤害吗？当野心存在时，爱还能存在吗？拜托，你必须面对这一切。但是你没有，因为你被困在了例行公事之中，困在了诸如性之类的感官享受的重复之中，等等。而爱与快感、与感官享受毫无关系，爱不是思想的产物。所以它并不在大脑的结构之中，它是某种完全在大脑之外的东西。大脑在本质和结构上就是感官享受和神经反应等的工具，而在只有感官享受的地方，爱是无法存在的。记忆不是爱。

We should also talk over together what is a religious life and what is religion, though it is a very complex question. Man has always sought, has inquired into, has longed for something beyond the physical, beyond the everyday existence of pain and sorrow and pleasure. He has always sought something beyond—first in the clouds, in the thunder as the voice of God. Then he worshipped trees and stones. The primitives still do; in the villages far away from these ugly, beastly towns they still worship stones, trees and small images. Man wants to find out if there is something sacred. And the priest comes along and says, "I will point out to you, I will show you"— just as the guru does. And there are the rituals of the Western priests, their repetition, and the worship of their particular image. And you too have your own images. Or you may not believe in any of that; you may say, "I am an atheist, I do not believe in God, I am a humanitarian."But man has always wanted to find out something that may be beyond time, beyond all thought.

我们也应该一起来谈谈宗教生活是什么，宗教是什么，尽管这是一个非常复杂的问题。人类一直在追寻、在探究、在渴望某种超越了物质、超越了有着痛苦、悲伤和快乐的日常生活的东西。他一直在寻求某种超越的东西——先是在云朵中，在被当作神的声音的雷声中。然后他膜拜树木和石头。原住民还在这么做；在远离这些丑陋、可怕的城镇的那些村庄里，他们依旧在膜拜石头、树木和小小的神像。人类希望发现是否存在某种神圣的东西，然后牧师过来了，说“我会指给你看的，我来展示给你”——就像古鲁所做的一样。而西方的牧师也有他们的仪式、他们反复诵念的词句，以及对他们特有的神像的膜拜。而你也有你自己的神像。或者，你也许不信仰其中的任何一个，你也许会说：“我是个无神论者，我不信神，我是个人道主义者。”但是人类一直希望找到某种也许超越了时间、超越了所有思想的东西。

So we are going to inquire, exercise our brains, our reason and our logic to find out what is religion, what is a religious life. Is a religious life possible in this modern world? Which does not mean becoming a monk or joining an organized group of monks. We will be able to find out for ourselves what is really, truly, a religious life only when we understand what religions actually are and put aside all that, and not belong to any religion, to any organized religion, to any guru, and not have any psychological or so-called spiritual authority. There is no spiritual authority whatsoever. That is one of the crimes that we have committed—we have invented the mediator between truth and ourselves.

所以我们要来探究，运用我们的大脑、我们的理性和我们的逻辑，来弄清楚宗教是什么，宗教生活是什么。在这个现代世界中有可能过一种宗教生活吗？这并不意味着成为一个僧侣或者加入一群有组织的团体。只有当我们懂得了如今的各派宗教究竟是什么，并把那一切都抛在一边，不属于任何宗教、任何有组织的宗教、任何古鲁，并且没有任何心理上或者所谓精神上的权威时，我们才能亲自发现真正的宗教生活是什么。精神上的权威无论如何都是不存在的。这是我们所犯下的罪行之一——我们发明了真理和我们自己之间的媒介。

So you begin to inquire into what is religion, and in the very process of that inquiry you are living a religious life, not at the end of it. In the very process of looking, watching, discussing, doubting, questioning, and having no belief or faith, you are already living a religious life. We are going to do that now.

于是你开始探询宗教是什么，而在这探询的过程中，你就过上了一种宗教生活，而不是在最后。就在看、观察、探讨、质问、怀疑及不抱任何信念、信仰的过程本身之中，你就已经过上了一种宗教生活。我们现在就要这么做。

We seem to lose all reason, all logic and sanity when it comes to religious matters. So we have to be logical, rational; we have to doubt, to question. All the things man has put together—the gods, the saviours, the gurus and their authority—all that is not religion. That is merely the assumption of authority by the few. Or you give them authority. Have you ever noticed that where there is disorder socially or politically, there comes a dictator, a ruler? Where there is disorder politically, religiously, or in our life, we create authority. You are responsible for the authority. And there are people who are only too willing to accept that authority.

当涉及宗教上的事情时，我们似乎就丧失了所有的理性、所有的逻辑和清明。所以我们必须保持逻辑性，保持理性；我们必须质问，必须怀疑。人类所拼凑的一切——神明、救世主、古鲁和他们的权威——所有这些都不是宗教。这只不过是少数人充当了权威，或者是你给了他们权威。你有没有注意到，哪里有社会上或政治上的混乱，哪里就会出现一个独裁者、一个统治者？只要政治上、宗教上或者我们的生活中存在混乱，我们就会树立起权威。你对权威负有不可推卸的责任。而世上有些人就是太愿意接受那种权威了。

So together we are going to look at what is religion. Where there is fear, man inevitably seeks something that will protect him, safeguard him, that will hold him in a sense of certainty, complete security, because he is basically frightened. Out of that fear, we invent gods, out of that fear we invent all the rituals, all the circus that goes on in the name of religion. All the temples in this country, all the churches and the mosques are put together by thought. You may say that there is direct revelation, but you never question, doubt that revelation; you accept it. And if one uses logic and reason, one sees that all the superstitions that one has accumulated is not religion. Obviously. Can you put all that aside to find out what is the nature of religion, of the mind that holds the quality of religious living?

所以我们一起来看看宗教是什么。当心存恐惧时，人类就必然会寻求某种可以保卫他、守护他的东西，能够给他带来确定感、彻底的安全感的东西，因为他骨子里是恐惧的。出于这种恐惧，我们发明了神明，出于这种恐惧，我们发明了所有的仪式，发明了以宗教之名上演的所有闹剧。这个国家中所有的庙宇、所有的教堂和清真寺都由思想所造。你也许会说那里有直接的启迪，但你从不质疑、从不怀疑那种启迪；你接受了它。然而，如果你运用逻辑和理性，你就会发现自己所积累的所有迷信都不是宗教。显然如此。你能抛开那一切，去发现宗教的本质是什么，能够容纳宗教生活的心灵具有怎样的品质吗？

Can we, as human beings who are frightened, not invent, not create illusions but face fear? Fear can disappear completely when you hold it, remain with it, not escape from it, when you give your whole attention to it. It is like a light being thrown on fear, a great, flashing light. Then that fear disappears completely. And when there is no fear there is no god, there are no rituals; all that becomes unnecessary, stupid. All that is irreligious. The things that thought has invented are irreligious, because thought is merely a material process based on experience, knowledge and memory. And thought invents the whole rigmarole, the whole structure of organized religions, which have no meaning at all. Can you put aside all that voluntarily without seeking a reward at the end of it? Will you do it? When you do that, then you begin to ask: What is religion, and is there something beyond all time and thought?

我们，作为恐惧的人类，能不能不发明、不捏造幻觉，而是面对恐惧？当你握着恐惧，与它共处，不逃避它，当你为它付出你全部的注意力时，恐惧就会彻底消失。那就像是有一束光，一道强烈的亮光照在了恐惧上，此时恐惧就会消失得无影无踪。而当没有了恐惧时，神明就不会存在，仪式也不会存在，那一切都会变得愚蠢、多余。那一切都不是宗教。思想发明的一切都不是宗教，因为思想只是一个基于经验、知识和记忆的物质过程。思想发明了组织化的宗教的整个烦琐的程序和复杂的结构，而这一切都毫无意义。你能自愿地抛开这一切，不寻求任何最终的回报吗？你愿意这么做吗？当你这么做的时候，你就会开始探询：宗教是什么，以及是否存在某种超越了所有时间和思想的东西？

You may ask that question but if thought invents something beyond, then it is still a material process. We have said that thought is a material process because it is sustained, nourished, in the brain cells. The speaker is not a scientist, but you can watch it in yourself, watch the activity of your own brain, which is the activity of thought. So if you can put aside all that easily, without any resistance, then you inevitably ask:“Is there something beyond all time and space? Is there something that has never before been seen by any man? Is there something immensely sacred? Is there something that the brain has never touched?"You will find that out if you have taken the first step—which is to wipe away all this thing called religion by using your brain, by your logic, your doubt, your questioning.

你也许会问这个问题，但是如果思想编造出了某种超越的东西，那么这依然是一个物质过程。我们说过思想是一个物质过程，因为它是在脑细胞中得到维系和滋养的。讲话者不是一个科学家，但是你能在自己身上观察这一点，观察你自己大脑的活动，也就是思想的活动。所以，如果你能够轻松地抛开那一切，毫无抗拒，那么你必然会问："是否存在某种超越了所有时间和空间的东西？是否存在某种任何前人都未曾见到的东西？是否存在某种无限神圣的东西？是否存在某种大脑从未触及的东西？"如果你迈出了第一步——

也就是运用你的大脑、你的逻辑、你的怀疑、你的质问抹掉了所有那些叫作宗教的东西——那么你就会发现真相。

Then what is meditation? That is part of so-called religion. What is meditation? Is it to escape from the noise of the world? To have a silent mind, a quiet mind, a peaceful mind? And you practise systems, methods, to become aware, to keep your thoughts under control. You sit cross-legged and repeat some mantra. I am told that the etymological meaning of that word“mantra”is“ponder over not—becoming”. That is one of the meanings. And it also means “absolve, put aside all self-centred activity”. That is the real, root meaning of mantra. But we repeat, repeat, repeat, and carry on with our self-interest, our egoistic ways, and so mantra has lost its meaning. So what is meditation? Is meditation a conscious effort?

We meditate consciously, practise in order to achieve something, to achieve a quiet mind, a sense of stillness of the brain. What is the difference between that meditator and the man who says"I want money, so I will work for it"? What is the difference between the two? Both are seeking an achievement, aren't they? One is called spiritual achievement, the other is called mundane achievement. But they are both in the line of achievement. So, to the speaker, that is not meditation at all. Any conscious, deliberate, active desire with its will is not meditation.

那么冥想是什么？这也是所谓宗教的一部分。冥想是什么？是逃避俗世的喧嚣吗？是拥有一颗安宁、寂静、平和的心吗？你修习各种体系、各种方法，希望变得觉知，希望控制你的思想。你盘腿坐下，反复诵念某些曼陀罗。有人告诉我“曼陀罗”这个词在语源上的意思是“深入思考‘不成为’”。这是其中一个含义。

它还意味着“消除、摒弃所有自我中心的行为”。这是曼陀罗真正的、根本的含义。但是我们反反复复地念念有词，同时继续着我们自私自利、自我中心的行为方式，所以曼陀罗失去了它原有的意义。那么冥想是什么？冥想是一种有意识的努力吗？我们有意识地冥想、练习，以期实现什么，获得一颗安静的心、头脑的一种宁静感。冥想者和说“我想发财，所以我会为此努力工作”的人有什么不同吗？这两者有什么不同吗？两者都在寻求一种成就，不是吗？一个被称为精神成就，另一个被称为世俗成就，但它们都属于成就的范畴。所以在讲话者看来，这根本不是冥想。任何有意识的、刻意的、主动的愿望，连同它的意志力，都不是冥想。

So one has to ask: Is there a meditation that is not brought about by thought? Is there a meditation which you are not aware of? Any deliberate process of meditation is not meditation; that is so obvious. You can sit cross-legged for the rest of your life, meditate, breathe and do all that business, but you will not come anywhere near the other thing. Because all that is a deliberate action to achieve a result—the cause and the effect. But the effect becomes the cause. So it is a cycle you are caught in. So is there a meditation that is not put together by desire, by will, by effort? The speaker says there is. You don't have to believe it. On the contrary, you must doubt it, question it as the speaker has questioned it, doubted it, torn it apart. Is there a meditation that is not contrived, organized? To go into that, we must understand the brain which is conditioned, limited. And that brain is trying to comprehend the limitless, the immeasurable, the timeless—if there is such a thing as the timeless.

所以你必须问一问：并非由思想引发的冥想是否存在？

有没有一种冥想是你没有觉察到的？任何刻意的冥想过程都不是冥想，这太显而易见了。你可以余生都盘腿坐着冥想、呼吸及去做诸如此类的把戏，但你无论如何都不会靠近那另一个事物。因为那一切都是为了实现某个结果的刻意的行动——这里就有原因和结果。但结果会变成原因，所以那就成了你所陷入的循环。那么，有没有一种冥想不是由欲望、意志和努力带来的？讲话者说有，但你不必相信这个说法。恰恰相反，你必须质问这一点，怀疑这一点，因为讲话者质问过了，怀疑过了，彻底检验过了。有没有一种冥想并不是虚构出来的、组织出来的？若要探究这一点，我们就必须了解深受制约的、局限的大脑。而那个大脑试图领悟无限的、不可衡量的、永恒的事物——如果存在永恒这回事的话。

And for that, it is important to understand sound. Sound and silence go together. You don't understand sound, the depth of sound, but you have separated sound from silence. Sound is the word, sound is your heart beating. The universe—universe in the sense of the whole earth, all the heavens, the million stars, the whole sky—is filled with sound. Obviously. You don't have to listen to scientists about it. And we have made that sound into something intolerable. So we want to have a brain that is quiet, peaceful. But when you listen to sound, the very listening is the silence. Silence and sound are not separate.

而正因为如此，所以说理解声音很重要。声音和寂静是并肩而行的。你不去理解声音、声音的深度，而是把声音和寂静分离了开来。声音就是词语，声音就是你的心跳。宇宙——宇宙是指整个地球、所有天体、万千星辰和整个天空——充满了声音。显然如此。你不必去听科学家对此说了什么。而我们把声音变成了某种无法忍受的东西，所以我们想拥有一颗寂静的、安宁的头脑。但是，当你聆听声音时，这种聆听本身就是寂静。寂静和声音是分不开的。

So meditation is something that is not contrived, organized. Meditation begins at the first step, which is to be free of all your psychological hurts, free of all your accumulated fears, anxieties, loneliness, despair, and sorrow. That is the foundation. That is the first step, and the first step is the last step. If you take that first step, it is over. But we are unwilling to take that first step because we don't want to be free. We want to depend—on power, on other people, on environment, on our experience and knowledge. You are always depending, depending, and are never free of all dependence, all fear. Therefore the ending of sorrow is love. Where there is that love there is compassion. And that compassion has its own integral intelligence. And when that intelligence acts, its action is always true. There is no conflict where there is that intelligence.

所以冥想是某种并非虚构出来的、组织出来的事情。冥想开始于第一步，也就是摆脱你所有的心理伤害，摆脱你积累起来的所有恐惧、焦虑、孤独、绝望和悲伤。这是基础，这是第一步，而第一步就是最后一步。如果你迈出了第一步，那就结束了。但是我们不愿意迈出第一步，因为我们不想获得自由。我们想要依赖——依赖权力、依赖别人、依赖环境、依赖我们的经验和知识。你一直在依赖、依赖，从来没有摆脱所有的依赖、所有的恐惧。所以悲伤的终结就是爱。哪里有这样的爱，哪里就有慈悲。而这慈悲有它自身完整的智慧。当那智慧行动时，它的行动始终都是正确的。那智慧存在时，冲突就不存在。

You have heard all this. You have heard about the ending of fear, the ending of sorrow, and about beauty and love. But the hearing is one thing, and action is another. You have heard all these things which are true, logical, sane, rational, but you won't act according to that. You will go home and begin all over again—your worries, your conflicts, your miseries. So one asks,"What is the point of it all? What is the point of listening to the speaker and not living it?"The listening and not doing it is a waste of your life. If you listen to something that is the truth and do not act, you are wasting your life. And life is much too precious. It is the only thing that we have. And we have also lost touch with nature, which means we have lost touch with ourselves, which is part of nature. You do not love the trees, the birds, the waters, the mountains. We are destroying the earth. And we are destroying each other. All that is such a waste of life. When one realizes all this, not merely intellectually or verbally, then one lives a religious life. To put on a loincloth or to go around begging or to join a monastery—

that is not a religious life. The religious life begins when there is no conflict, when there is this sense of love. Then you can love another, your wife or your husband, but that love is shared by all human beings. It is not given to one person and is therefore not restricted.

你听到了这一切。你听到了恐惧的终结、悲伤的终结，也听到了关于美和爱的一切。但听到是一回事，行动是另外一回事。你听到了所有这些真实的、合理的、符合逻辑和理性的事情，但你不愿意这样去行动。你回到家之后又会重蹈覆辙——你的忧虑，你的冲突，你的痛苦依然如故。所以我要问："这一切意义何在？听讲话者讲这些然后不把它活出来，这意义又何在？"只听不做就是在浪费你的生命。如果你听到了某种真实的东西却不去行动，你就是在浪费自己的生命。但生命实在太宝贵了，这是我们唯一拥有的东西。同时我们也失去了与自然的联系，这就意味着我们失去了与我们自己的联系，因为我们就是自然的一部分。我们不爱树木、飞鸟、流水和群山。我们在摧毁地球，我们也在互相摧毁。这一切真是一种对生命的浪费。当你认识到了这一切，而不是仅仅从智力上或者字面上理解了，那么你就过上了宗教生活。当冲突不复存在，当有了这种爱时，宗教生活就开始了。然后你就可以去爱别人，去爱你的妻子或丈夫了，但那爱是为所有人类所共享的。它不是只给予一个人的，因此是没有局限的。

So, if you give your heart and mind and brain, there is something that is beyond all time. And there is the benediction of that. Not in temples, not in churches, not in mosques. All Possible Life.

所以，如果你付出你全部的身心和头脑，你就会发现有一种超越了所有时间的事物，此时就会有那样一种至福。它不在庙宇中，不在教堂中，不在清真寺中，那至福就在你身边。

一九八五年二月十日

图书在版编目（CIP）数据

生命的所有可能：汉、英 /（印）克里希那穆提著；王晓霞译 . -- 北京：北京时代华文书局，2022.4

书名原文：That Benediction is Where You Are

ISBN 978-7-5699-3738-1

Ⅰ . ①生… Ⅱ . ①克… ②王… Ⅲ . ①人生哲学－通俗读物－汉、英 Ⅳ . ① B821-49

中国版本图书馆 CIP 数据核字 (2020) 第 096322 号

北京市版权局著作权合同登记号　图字：01-2020-2752

Krishnamurti Foundation of America
P.O. Box 1560, Ojai, California 93024 USA
E-mail: kfa@kfa.org Website: www.kfa.org

生命的所有可能：汉、英

SHENGMING DE SUOYOU KENENG: HAN、YING

著　　者 | [印] 克里希那穆提
译　　者 | 王晓霞

出 版 人 | 陈　涛
选题策划 | 刘昭远
责任编辑 | 周海燕
责任校对 | 张彦翔
装帧设计 | 柒拾叁号
责任印制 | 訾　敬

出版发行 | 北京时代华文书局 http://www.bjsdsj.com.cn
北京市东城区安定门外大街 138 号皇城国际大厦 A 座 8 楼
邮编：100011　电话：010 - 83670692　64267677
印　　刷 | 北京盛通印刷股份有限公司　010-83670070
（如发现印装质量问题，请与印刷厂联系调换）
开　　本 | 787mm×1092mm　1/32　印　　张 | 7　字　　数 | 120 千字
版　　次 | 2022 年 4 月第 1 版　印　　次 | 2022 年 4 月第 1 次印刷
书　　号 | ISBN 978-7-5699-3738-1

定　　价 | 52.00 元